USE OF THE DUAL-TRACE
OSCILLOSCOPE
A Programmed Text

USE OF THE DUAL-TRACE OSCILLOSCOPE
A Programmed Text

CHARLES H. ROTH, JR.

Department of Electrical Engineering
The University of Texas at Austin

PRENTICE-HALL, INC., *Englewood Cliffs, NJ 07632*

Library of Congress Cataloging in Publication Data

Roth, Charles H.
  Use of the dual-trace oscilloscope.

  1. Cathode ray oscilloscope—Programmed instruction.
I. Title.
TK7878.7.R67   621.3815'48   81-17723
                                   AACR2
ISBN 0-13-940031-1
ISBN 0-13-940023-0 (pbk)

Printed in the United States of America

10

ISBN 0-13-940031-1

ISBN 0-13-940023-0 {pbk}

Prentice-Hall International (UK) Limited, *London*
Prentice-Hall of Australia Pty. Limited, *Sydney*
Prentice-Hall Canada Inc., *Toronto*
Prentice-Hall Hispanoamericana, S.A., *Mexico*
Prentice-Hall of India Private Limited, *New Delhi*
Prentice-Hall of Japan, Inc., *Tokyo*
Simon & Schuster Asia Pte. Ltd., *Signapore*
Editora Prentice-Hall do Brasil, Ltda., *Rio de Janeiro*

CONTENTS

PREFACE

This programmed text has been designed for use in a basic electrical engineering laboratory course. Since the oscilloscope is one of the most important and versatile measuring instruments, instruction in use of the oscilloscope is usually given near the beginning of such a course. The modern triggered sweep oscilloscope is a rather complex instrument with many controls and input terminals. The instruction manuals supplied with most scopes are intended for reference and are usually not suitable for teaching an inexperienced student how to operate a scope. This programmed text breaks down the process of operating a scope into a series of logical steps starting with deflection of the electron beam and continuing through proper use of the triggering controls to measure the phase difference between two waveforms. Understanding the operation of the scope is stressed at each step rather than mere manipulation of controls. Although teaching use of the scope is the primary purpose of this text, the student will also learn some general principles of electrical measurements which are applicable to other laboratory instruments.

Programmed Instruction

Programmed instruction is especially well suited to teaching the effective use of laboratory equipment. The material to be learned is broken down into a carefully designed sequence of steps or frames. Each frame explains a basic idea or presents a problem to be solved. The student then must take some positive action, which usually includes writing an answer. He thus participates actively in the learning process and "learns by doing." The correct answer is provided on the next page so that the student can immediately verify his own answer. This provides a reinforcement if the answer is correct, or correction and explanation if it is wrong. Each student can thus proceed at his own pace, going on to a new set of material only after he has mastered the previous set.

By careful study of student responses and examination results, this program has been revised and improved to remove "stumbling blocks" to student progress. The size of the steps in the program has been adjusted so that the average student will get more than 90 per cent of the answers correct as he progresses through the program. This does not imply that the program is trivial; the student is expected to reason out many things for himself. For example, he must draw his own circuit diagrams rather than being told how to hook up the circuit. Our experience with this programmed text indicates that every student who has the necessary prerequisites and who completes the program according to instructions will be able to demonstrate mastery of the material by passing a practical examination on use of the scope.

Prerequisites

This program assumes a knowledge of basic DC and AC circuits. Familiarity with circuit terms and the ability to solve simple circuits is required. No specific previous laboratory experience is assumed, but the laboratory parts of the program require the ability to hook up simple circuits and read meters.

Required Equipment

This text is intended for use with a dual-trace, triggered-sweep oscilloscope such as the Tektronix 922 or Tektronix 2213. A table of suitable oscilloscopes is given in Appendix B. The theory parts of the text are general enough to be used with almost any triggered sweep oscilloscope, but the laboratory parts will require some modification to adapt them to different scopes. When scopes other than the Tektronix 922 or 2213 are used, the laboratory instructor should prepare a list of any changes which are required for the particular scope being used. Suggestions for adapting the program to other types of scopes are given in the Teacher's Manual. Other required equipment includes oscillators, function generators, and DC power supplies which are readily available in most laboratories. Since detailed instructions for using this auxiliary equipment are not included in the program, the lab instructor should demonstrate the use of this equipment as required.

Acknowledgements

Preparation of the earlier versions of this text was supported in part by the American Society for Engineering Education Programmed Learning Project which was sponsored principally by the Ford Foundation and directed by Norman Balabanian. Special thanks are due to A. A. Root, editor for the Programmed Learning Project, for his criticism and helpful suggestions. Many lab instructors and students worked with the different versions of this program and helped to provide data for the necessary revisions. Their contribution is gratefully acknowledged.

C. H. Roth, Jr.
The University of Texas at Austin

INTRODUCTION

This text is divided into nine parts. Four of these parts are for preparation outside of the laboratory, and four parts are used in the lab working with the oscilloscope. In Preparation Part I and Laboratory Part I, you will learn the relation between voltages applied at the scope terminals and the pattern which will be displayed on the screen. Part II covers displaying waveforms on the scope as a function of time. You will learn how to make accurate measurements with the scope in Preparation Part III and Lab Part III. Measurement of the phase angle between sinusoidal voltages is covered in Part IV. Part V is intended for review after you have completed the programmed instruction in the other parts.

Objectives

1. When you complete this programmed text you should be able to use the scope effectively in future laboratory work. Effective use of the scope includes the ability to:

   (a) Avoid damage to the scope.
   (b) Adjust the scope and check its calibration.
   (c) Connect the scope to the circuit with minimum disturbance to the quantity being observed.
   (d) Display a waveform (or a selected portion thereof) and measure its characteristics (amplitude, period, frequency).
   (e) Determine the relationships between two waveforms such as phase shift.
   (f) Display x-y plots.
   (g) Interpret the results of oscilloscope measurements taking the limitations of the scope into account.

2. You should understand the operation of the scope in the sense that you are able to:

   (a) Explain the functions of the controls and input terminals and their relationship to the trace (using a simple block diagram if necessary).
   (b) Given the input signals and control settings, sketch the sweep waveform which is generated within the scope.
   (c) Given the input signals and control settings, predict the trace which will appear on the screen.
   (d) Given a desired trace, determine the necessary input signals and control settings.

<u>Read the following instructions carefully before proceeding:</u>

1. This programmed text is divided into frames. Each frame will ask a question or require you to take some action. The answer to each frame is given on the following page.

2. Read each frame carefully. Then write your answer(s) in the space provided or take the appropriate action. If you cannot answer the question, leave the space blank.

3. After you have written your answer(s), turn the page and check your answer(s). Most of the time you will be correct.

4. If your answer is wrong or if you left it blank, go back and correct it. Mark the incorrect frame with an X so you can review it later. Do not go on until you understand why your answer was wrong; ask your instructor for help if necessary.

5. Continue through the program frame by frame as described above. After completing each part, go back and review any frames which you missed the first time through.

6. Avoid the temptation to look at the correct answer before your write your own answer. Effective learning will occur only if you carefully study each frame and then write your own answer.

7. You will <u>not</u> be graded on your answers. You will "cheat" only yourself if you copy the answers instead of working them out.

8. This program is intended for <u>individual</u> instruction. Each student should schedule time on the oscilloscope so that he can work alone.

9. Each part is divided into several sections as indicated in the table of contents for that part. Plan your study time so that you can complete a given section at one sitting. Take your breaks in between sections. For example, for Preparation Part I, the best time to take a break is after frame 1.24 or 1.38.

10. To complete the program, you must work through the following pages four times--once for each of the four parts. Frames for Part I are found on the top half of the right-hand pages; frames for Part II on the bottom half. Similarly, frames for Parts III and IV are at the top and bottom of the left-hand pages.

11. The preparation for each part should be completed before going to lab. Before you start the lab work, your lab instructor or supervisor will check to see that you have completed the preparation and will answer any questions which you might have.

PREPARATION   PART I

1.1 The oscilloscope is an electronic measuring instrument which displays electrical signals in graphic form. It is probably the most widely-used electronic instrument because it can be used to observe waveforms as well as
measure voltage, time, frequency, and phase angle.

The purpose of this part is to prepare you for basic operation of the
oscilloscope with external voltages applied to both the vertical and horizontal inputs. When you complete this part you should know the relation
between voltages applied at the scope terminals and the pattern which will
be displayed on the screen. You must complete this preparation <u>before</u> you
begin the related laboratory work.

Before proceeding, read the instructions on p. 2 if you have not already
done so.

TURN TO FRAME 1.2 ON PAGE 5

PREPARATION   PART II

2.1 In this part you will learn how to display waveforms on the scope as a
function of time.

<u>Objectives</u>

1. Explain why a periodic signal is necessary in order to obtain a steady
display on the screen.

2. State the two main parts of the TIME BASE and explain their functions.

3. Sketch the sweep waveform, showing its relation to the trigger waveform
and the waveform displayed on the screen.

4. Explain the difference between TRIGGER SLOPE and TRIGGER LEVEL. Explain the function of each.

5. Explain the options available on the TRIGGER SOURCE.

6. Explain the difference between AC and DC trigger coupling.

7. Explain the difference between AUTO and NORMal triggering.

8. Explain dual-trace operation, including the difference between chop and
alternate modes.

3

PREPARATION PART III

3.1 After completing this part you should know the techniques required to make accurate measurements with the scope. You should understand the effects of loading, and the use of the probe to reduce loading.

Objectives

1. Explain how loading affects the accuracy of electrical measurements, and be able to calculate the error due to loading.

2. Explain how a probe can be used to reduce loading error. Give the equivalent circuit for the probe and scope input. Explain how the probe can be adjusted so that the attenuation is independent of frequency. Calculate the input voltage to the probe given the vertical deflection and vertical sensitivity setting.

3. Explain the various types of errors which can occur when using the scope, and explain how to reduce the effects of these errors.

4. Explain the overall purpose and operation of the differential amplifier mode. Show how to connect the differential amplifier. Given the input signals to the differential amplifier, the sensitivity setting and the common mode rejection ratio, determine the trace which will appear on the screen.

PREPARATION PART IV

4.1 Upon completion of this part you should be prepared to measure the phase angle between sinusoidal voltages by the dual-trace, triggered sweep and ellipse methods.

Objectives

1. Explain how to measure the phase angle between two sinusoidal voltages using the dual-trace method. Indicate proper connections to the scope and proper settings for the scope controls. Be able to determine the phase angle from the scope display.

2. Same as (1) using the triggered sweep method.

3. Explain how to measure the phase angle by the ellipse method including connections to the scope, setting the scope controls, and determination of the phase angle from the ellipse. Also explain how the Webb mask is used.

4. For the dual-trace, triggered sweep and ellipse methods, sketch the expected scope display given the sinusoidal input voltages.

5. Determine the frequency ratio of two sine waves from a Lissajous figure.

4

1.2 The objectives for this part are:

1. Explain how to avoid damage to the scope screen.

2. Show how to connect external signal sources to the scope taking proper grounding into account.

3. Given a periodic signal, determine the AC and DC components of that signal.

4. Given the voltages applied to the vertical and horizontal inputs of the scope and vertical and horizontal sensitivity settings, determine the pattern which will appear on the screen.

5. Given a desired trace on the scope screen, determine the necessary vertical and horizontal input voltages and sensitivity settings.

2.2 <u>Displaying Waveforms as a Function of Time</u>

In this and the following frames, assume that the vertical and horizontal sensitivities are $S_v = S_h = 1$ volt/div. Also assume that the spot is centered at the left edge of the grid when no inputs are present. Thus the point $x = 0$, $y = 0$ will be at the center of the left edge of the grid.

It is desired that the spot move at a uniform rate from point A to point B in 4 seconds. Plot the required input voltages v_v and v_h as functions of time.

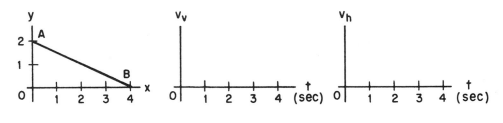

FIG. 2-2

In order to use the scope to make accurate measurements, the scope must be properly calibrated, and the circuit being tested must not be affected when the scope is connected to it. We will now consider the effect of measuring instruments on the quantity being measured.

Measurement Errors Due to Loading

An <u>ideal</u> measuring instrument does not disturb the circuit to which it is connected in any way. In other words, all the voltages and currents in a circuit are the same regardless of whether the ideal instrument is connected to the circuit or not.

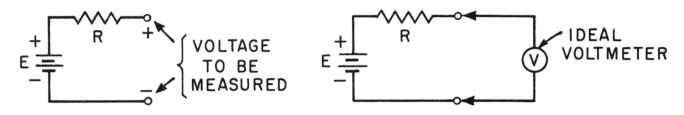

If an <u>ideal</u> voltmeter is connected as shown, the current flowing in

the circuit is _____. The ideal voltmeter will read   (a)

_____. The input impedance of the ideal voltmeter   (b)

is _____.   (c)

4.2 Given two sinusoidal voltages, we can always choose the time origin so that the phase angle associated with one of them is 0. Therefore, we will assume that the given voltages are of the form

$$v_1 = A \sin \omega t \quad \text{and} \quad v_2 = B \sin (\omega t + \theta)$$

The voltage with 0 phase angle (v_1) will be referred to as the reference voltage. The phase angle of v_2, θ, is then the phase difference between the two voltages. Our problem is to measure θ. We will first review how to determine θ by inspection of a plot of v_1 and v_2.

<u>Deflection of the Electron Beam</u>

The oscilloscope displays electrical signals on the screen of a <u>cathode-ray tube</u>. (The picture tube in a TV set is one type of cathode-ray tube.) As shown in Fig. 1-3, the evacuated glass envelope of the cathode-ray tube contains a(an) _____ _____ (a) which produces an electron beam, and _____ _____ (b) which can be used to deflect the electron beam.

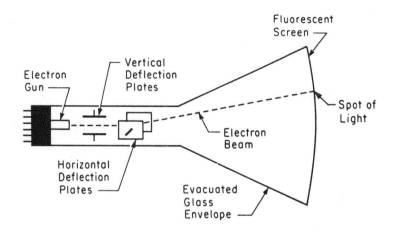

FIG. 1-3. CATHODE RAY TUBE

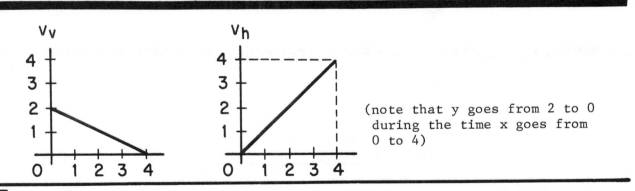

(note that y goes from 2 to 0 during the time x goes from 0 to 4)

2.3 After reaching point B (Fig. 2-2), we want the spot to jump back to point A and then retrace the line from A to B at the same rate as before. This action should be repeated twice more (four times in all). Sketch the required input voltages v_v and v_h.

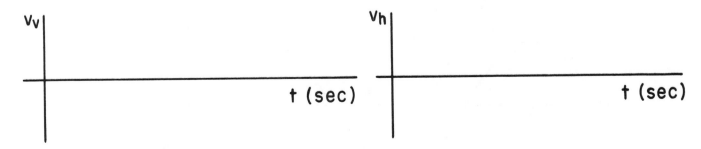

7

(a) zero   (b) E   (c) infinite

(Since no current was flowing before connecting the meter, we want no
current to flow after connecting the meter. Therefore the ideal meter
must act like an open circuit and have an infinite input impedance.)

3.3 A non-ideal voltmeter has a finite input impedance. It will draw
some current from the circuit being tested and may change the voltage
being measured.

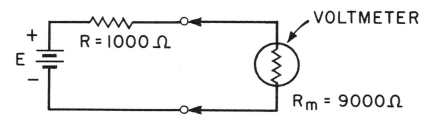

(R$_m$ represents the
input impedance of
the voltmeter.)

In the above circuit, the voltmeter will read _____. (a)

The voltmeter reading is less than the correct value because the

voltmeter loads down the circuit. The per cent error in the voltage

reading due to loading is _____. The error due to loading   (b)

would be negligible only if the impedance of the voltmeter, R$_m$, is

_____ than R.   (c)

Determination of Phase Angle

4.3 The plot below shows $v_2 = B \sin(\omega t + \theta)$. From the equation,

at t = 0, v_2 = _____.   (a)

— If t = 0 here, θ = _____ degrees.   (b)

— If t = 0 here, θ = _____ degrees.   (c)

— If t = 0 here, θ = _____ degrees.   (d)

— If t = 0 here, θ = _____ degrees.   (e)

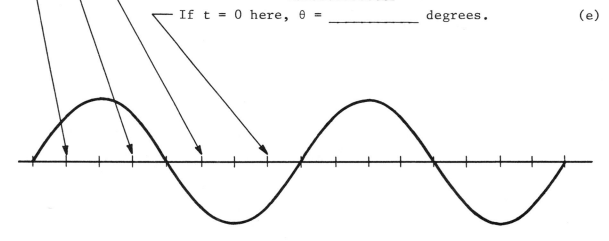

Answers to frame 1.3:　　(a)　electron gun
　　　　　　　　　　　　　(b)　(vertical and horizontal) deflection
　　　　　　　　　　　　　　　　plates

| 1.4 | The screen is coated with a phosphor which produces a spot of light

when the _____ _____ strikes it. Varying the　　　　(a)

intensity of the electron beam will cause a corresponding variation

in the intensity of the _____.　　　　　　　　　　　　　　(b)

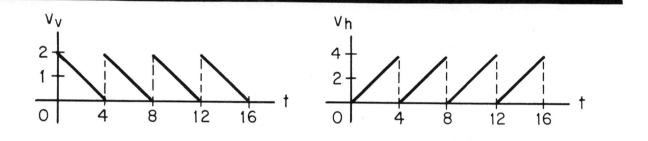

| 2.4 | In the frames which follow, still assume that the spot is centered at the
left edge of the grid when no inputs are present and $S_v = S_h = 1$ volt/div.

It is desired that the spot trace out
a triangle on the screen as shown. The
position of the spot at t = 0, 1 ms,
2 ms, and 3 ms is indicated. Plot the
required input voltages v_v and v_h as a
function of time.

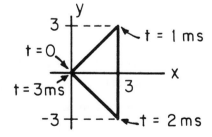

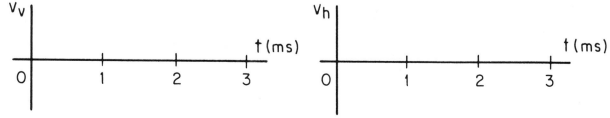

(a) .9E   (b) 10%   (c) much greater

3.4

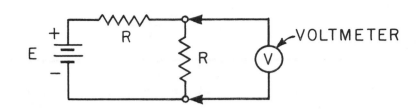

If the meter is ideal, it will read _____.   (a)

If the meter is non-ideal it will load down the circuit. When cal-
culating the effect of loading on a circuit, first replace the circuit seen at the meter terminals by its Thevenin's equivalent:

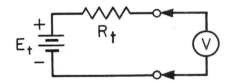

E_t = _____   (b)

R_t = _____   (c)

If the meter has an input impedance R_m it will read _____   (d)
(express your answer in terms of E_t, R_t and R_m)

(a) v_2 = B sin θ (b) 45   (c) 135   (d) −135 (or + 225)
(e) −45 (or +315)

4.4   In this and the following frames, when we write an expression of the
form v_2 = B sin ($\omega t + \theta$) we will assume that B is positive and we
will choose the magnitude of θ to be less than or equal to 180°.
Thus instead of writing θ = 280°, we will use the equivalent value
θ = _____, and instead of 190° we will use   (a)

_____.   (b)

Given: $v_2(t)$ = B sin ($\omega t + \theta$)

If v_2 is positive at t = 0, then sign of θ is _____.   (c)

If v_2 is negative at t = 0, the sign of θ is _____.   (d)

Given a plot of $v_2(t)$, how can we tell the sign of θ?

_____

_____   (e)

10

(a) electron beam
(b) spot (of light)

1.5   When the electron beam strikes the screen, it produces heat as well
       as light. If a high-intensity spot is left stationary on the screen
       for a period of time, the screen will be overheated and a burned spot
       will result. This causes permanent damage to the cathode-ray tube.
       Damage to the cathode-ray tube screen may be avoided by

           keeping the intensity _____ when the spot is stationary   (a)
           Keeping the spot _____ when the intensity is high.   (b)

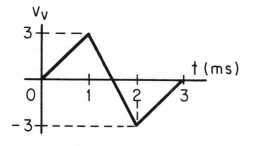   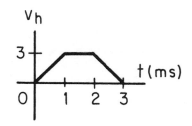

2.5   The picture traced out in frame 2.4 would fade out as soon as the input
       signals were over. In order to observe a steady picture on the scope
       screen, it would be necessary to retrace the figure on the screen periodic-
       ally. This could be accomplished by _____

       _____

(a)   E/2   (b)   E/2   (c)   R/2

(d) $E_t \dfrac{R_m}{R_m + R_t}$

3.5   A voltmeter is used to measure the output voltage of a network as shown.

The Thevenin's equivalent of the network is

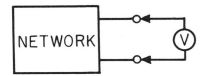
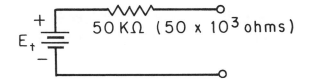

The meter reads 9.5 volts and has an input impedance of 1 megohm (10^6 ohms).

What would an ideal voltmeter read? _____   (a)

What is the per cent error due to loading? _____   (b)

(Per cent error is defined as $\dfrac{\text{error}}{\text{correct value}}$ x 100%.)

(a) -80° (b) -170° (c)   positive   (d)   negative

(e)   The sign of θ is the same as the sign of v_2 at t = 0.

4.5 $v_2(t) = B \sin(\omega t + \theta)$ is plotted below for two values of θ.

In both cases, the sign of θ is _____.   (a)

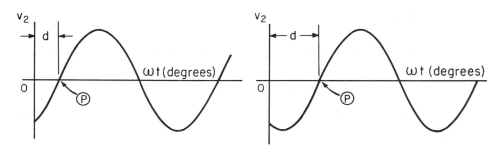

The "0° point" on the sine wave (that point at which a full cycle of the sine wave starts) is labeled P. At point P, $\omega t + \theta =$ _____   (b)
so $\theta =$ _____.   (c)

For the case where θ is negative, if the distance between the origin and the "0° point" on the sine wave is d (in degrees),
$\theta =$ _____   (d)

(a)   low (turned down)
(b)   moving (in motion)

1.6 When no voltage is applied to either set of deflection plates, the electron beam passes straight through between the plates and produces a spot on the center of the screen.

Now consider the effect of applying a voltage between the vertical deflection plates. If the top plate (see below) is made positive with respect to the bottom plate, the negatively-charged electrons are attracted toward the (top/bottom) _____ plate and the   (a)
beam is deflected (upward/downward) _____   (b)

If the right horizontal deflection plate is made positive with respect to the left plate, the spot on the screen will be deflected to the _____.   (c)

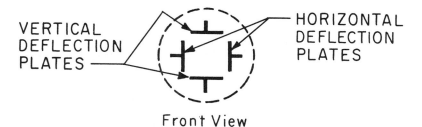

Front View

---

repeating the input signals periodically

2.6 The following waveform is applied to the vertical input:

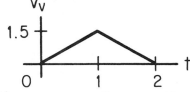

We wish to trace out the same waveform on the scope face:

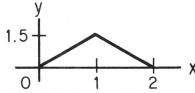

Sketch the waveform which must be applied to the horizontal amplifier input.

V_h

$\begin{array}{c|c c c}\hline & & & \\ \hline 0 & 1 & 2 & 3 \end{array}$ t

13

(a) $$\frac{10^6 E_t}{50 \times 10^3 + 10^6} = 9.5 \qquad E_t = 9.5 \times 1.05 \approx \underline{10}$$

(b) $$\frac{10 - 9.5}{10} \times 100\% = \underline{5\%}$$

3.6   Now consider a measuring instrument whose input circuit is a capacitor.

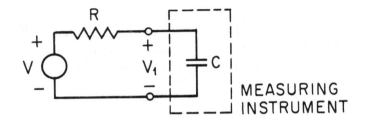

MEASURING INSTRUMENT

If V is a DC source, the capacitor acts like a(n) _____ cir-   (a)

cuit, so the input voltage to the measuring instrument is _____.   (b)

If V is an AC source, for very high frequencies the capacitor ap-

proaches a(n) _____ circuit, so the input voltage to the   (c)

measuring instrument approaches _____.   (d)

(a) negative   (b) 0   (c) $-\omega t$ (d) $-d$

4.6 $v_2 = B \sin(\omega t + \theta)$ is plotted below for two values of θ.
In each case, give the value of θ.

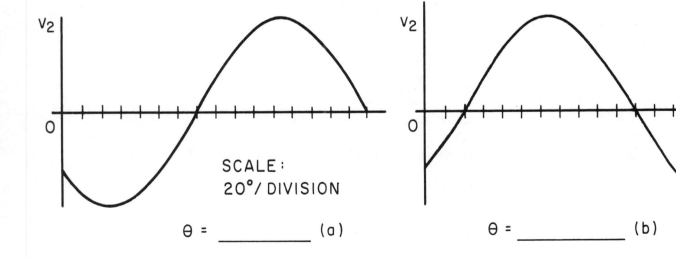

SCALE:
20°/DIVISION

$\theta = $ _____ (a) $\theta = $ _____ (b)

1.7 When the input signals are connected directly to the deflection plates, a large voltage is required to deflect the spot a small amount (in the order of 20 volts for one centimeter deflection). In order to observe small signals, these signals must be amplified before being applied to the deflection plates as shown in Fig. 1-7. Each amplifier multiplies its input voltage by a <u>positive</u> constant.

In Fig. 1-7, if v_v is positive the _____ plate will be positive   (a)

with respect to the _____ plate.   (b)

If v_h is positive the _____ plate will be positive with respect   (c)

to the _____ plate.   (d)

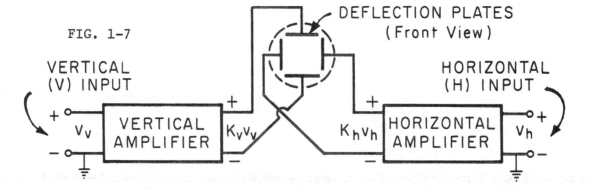

FIG. 1-7

VERTICAL (V) INPUT

HORIZONTAL (H) INPUT

DEFLECTION PLATES (Front View)

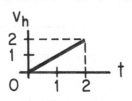

2.7 If the vertical input is periodic and we wish to observe a single period on the scope, we could trace out a single period of the waveform only once on the scope screen; however, this picture would fade out immediately and would be difficult to observe. We can obtain a steady picture by <u>retracing</u> the same waveform over and over again. To do this, we must apply a periodic sweep waveform to the horizontal amplifier input. If v_v and v_h are as shown, the maximum horizontal

deflection is _____ div. Sketch the trace which appears on   (a)
the screen. (S_v and S_h are still 1 volt/div.)

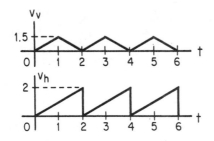

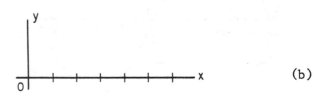

(b)

(a)   open   (b)   V   (c)   short   (d)   0

3.7   If V is an AC source, the loading
effect of the capacitor on V_1 will
depend on the frequency.

In terms of ω and C, the magnitude
of the impedance of the capacitor
is _____ (a)

Therefore, as the frequency increases, the impedance of the
capacitor will (increase/
decrease) _____ (b)

and the voltage measured by the
measuring instrument will
_____. (c)

If the magnitude of V remains
constant, make a rough sketch of
the magnitude of V_1 as a function of frequency.

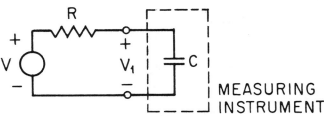

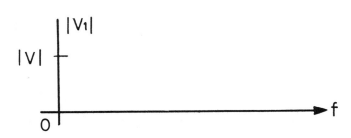

(a) -140° (b) -40°

4.7   Now consider the case where the phase angle is positive. If the waveform starts at t = 0, the "0° point" on the sine wave will not be
visible. Imagine the sine wave extended back to where it crosses the
ωt axis as shown by the dotted lines on the plots below. If point P
were visible, we could read the value of θ directly off the plot as
indicated. Since P will not be visible, we must compute θ indirectly.
The "180° point" on the sine wave is labeled Q. In terms of θ, the
distance between the origin and Q is _____.   (a)

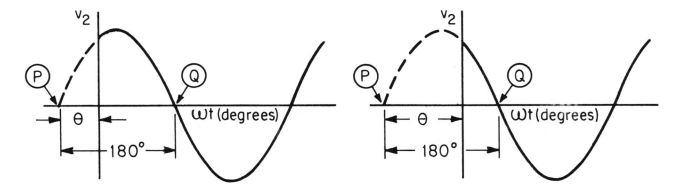

For the case where θ is positive, if the distance between the origin
and the "180° point" on the sine wave is d (in degrees), θ = _____. (b)

(a) top   (b) bottom   (c) right   (d) left
     (upper)   (lower)

1.8 In Fig. 1-7, if the vertical input voltage is negative, the spot on
     the screen will be deflected _____. To deflect the spot   (a)
     to the right, a _____ voltage must be applied to the   (b)
     _____ input.   (c)

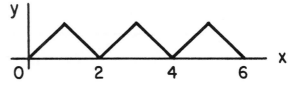

If your answer to (b) is [graph] continue

with this frame; otherwise, turn to frame 2.9.

2.8 In frame 2.7 the horizontal signal is

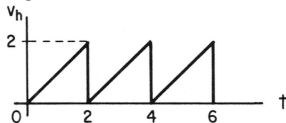

Since the peak value of v_h is 2, and we have assumed a horizontal sensitivity of 1 volt/div., the maximum horizontal deflection is _____ div.
Now, the maximum horizontal deflection in your answer is 6 div., so it
can't be right. Go back to frame 2.7 and try again.

(a) $1/\omega C$

(b) decrease

(c) decrease

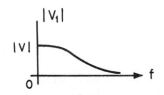

3.8 We will now consider the loading effect of the scope on the circuit being measured. When the input switch is set to DC, the input impedance of the vertical amplifier of the scope can be represented by

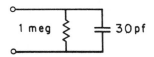

For DC voltages, the input impedance is _____ (a)

As the frequency is increased, the input impedance will _____ (b)
because of the shunt capacitance.

At very high frequencies, the impedance of the shunt capacitor becomes much less than 1 megohm, and practically all of the input current will flow through the capacitor. At such high frequencies, the
input circuit of the scope amplifier can be approximately represented
by a _____ of numerical value (c)
_____. (d)

(a) $180^{\circ} - \theta$ (b) $180^{\circ} - d$

4.8 For each of the following plots, give the value of the phase angle:

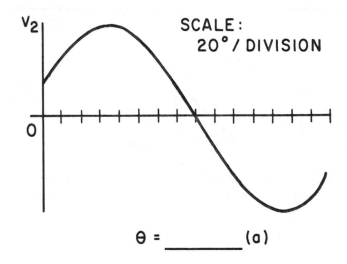

SCALE:
20°/DIVISION

$\theta =$ _____ (a)

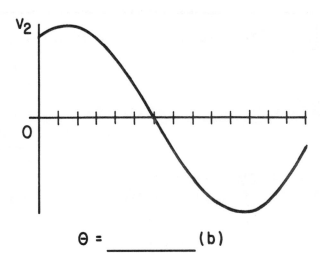

$\theta =$ _____ (b)

(a) downward   (b) positive   (c) horizontal

1.9 The inputs to the vertical and horizontal amplifiers (labeled v_v and v_h in Fig. 1-7) are brought out to terminals on the front of the scope as shown in Fig. 1-9. Voltage sources are connected to these input terminals with the polarities shown. The spot will appear in the _____ quadrant of the scope screen. (Remember that the polarity markings on v_v and v_h are reference polarities and do not necessarily correspond to the actual polarity of the applied voltage.)

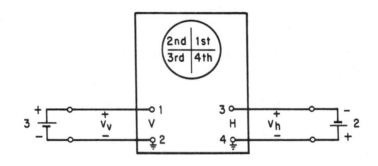

FIG. 1-9

Answers to 2.7.

(a) 2 div.   (b)

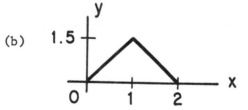

(Note that the waveform is retraced 3 times; it is not repeated.)

2.9 Generating the Sweep Waveform

If a signal $v_v(t)$ is applied to the vertical scope input and we wish to display v_v as a function of time, we must provide a horizontal signal of the type discussed in the preceding frames. The required horizontal signal is generated within the scope by the TIME BASE.

922 ONLY: Review Fig. 1-67 (p. 137)

CAL X-Y ONLY: Review Fig. 1-68 (p. 139)

If we wish to display v_v as a function of time, indicate the proper position of the horizontal amplifier input switch in Fig. 1-67 or 1-68. (a)

The two main parts of the TIME BASE are the TRIGGER PULSE GENERATOR and the SWEEP GENERATOR as shown in Fig. 5-4 (p. 235). When the scope is used to display a waveform as a function of time the output of the sweep generator is connected to the _____ amplifier.   (b)

Is an external horizontal (X) input used in this mode? _____   (c)

(a) 1 megohm   (b)   decreases

(c) capacitor (the resistor can be omitted since practically all of the input current flows through the capacitor at high frequencies)

(d) 30 pf

3.9 Draw an equivalent circuit which represents the input impedance of the vertical amplifier of the scope. Specify values for both components.

(a)

Draw an equivalent circuit which represents the input impedance of the scope for DC voltages.

(b)

Draw an equivalent circuit which approximately represents the input impedance of the scope for very high frequencies.

(c)

(a) 20° (b) 60°

4.9 Sketch the following waveforms:

(a) $v_2 = 10 \sin (\omega t + 140^{o})$

(b) $v_3 = 10 \sin (\omega t - 140^{o})$

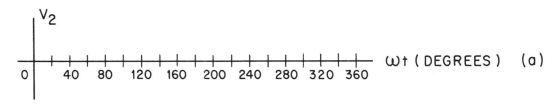

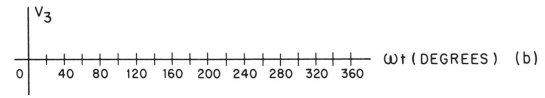

Check to see that your waveforms cross the ωt axis when $(\omega t + \theta) = 0, 180^{o}$, 360°, etc. Also check to see that your plots have the proper sign at $t = 0$.

2nd (since v_v is + and v_h is -)

Grounding

1.10 The ground symbol ($\perp$) which appears in Figs. 1-7 and 1-9 indicates that one vertical input terminal and one horizontal input terminal are "grounded," i.e., are connected to the scope case and chassis. (The chassis has the electronic components mounted on it and serves as a common electrical connection for many of the scope circuits.)

On Fig. 1-9 draw a line between two terminals which are connected directly together through the scope case and chassis. (a)

Give the voltage between each of the following terminal pairs in Fig. 1-9:

V_{12} = _____ V_{34} = _____ V_{24} = _____ (bcd)

V_{13} = _____ V_{14} = _____ V_{32} = _____ (efg)

NOTE: V_{12} means the voltage drop from 1 to 2, etc.

(a) switch is connected to TIME BASE (b) horizontal

(c) no

2.10 The SWEEP GENERATOR output is 0 until it is triggered by a pulse at the input. If a trigger pulse occurs, the following sweep waveform is generated at the output:

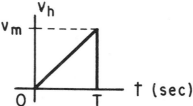

Assume that the peak voltage, v_m, is just sufficient to deflect the spot the full width of the screen. If a single trigger pulse occurs at the input of the sweep generator, describe the horizontal motion of the spot on the screen. (Include the time T in your answer.)

_____

_____

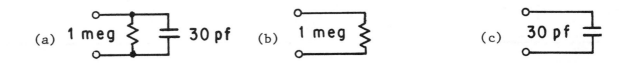

(a) 1 meg ⊰ ⊣⊢ 30 pf   (b) 1 meg ⊰   (c) 30 pf ⊣⊢

3.10   When the vertical amplifier input switch is set to AC, there is a blocking capacitor in series with the vertical amplifier input. However, the effect of this blocking capacitor on the input impedance is negligible except at very low frequencies.

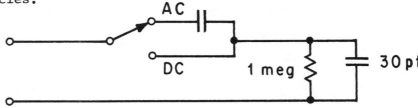

Over the frequency range where the AC input is normally used, should there be any significant difference between the input impedance with the switch set to AC and with the switch set to DC? _____

Why? _____

_____

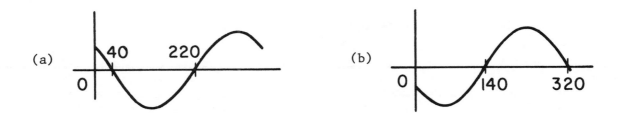

(a)   40   220   0

(b)   0   140   320

4.10   If $v_2 = 5 \sin(\omega t + \theta)$, complete the following table:

range of θ	$0° < \theta < 90°$	$90° < \theta < 180°$	$-180° < \theta < -90°$	$-90° < \theta < 0°$	
sign of $v_2(0)$					(a)
sketch of v_2					(b)

In which case is there a positive peak of the sine wave between the

origin and the first time the curve crosses the time axis? _____ (c)

A negative peak? _____ (d)

22

(a)

(b) $V_{12} = +3$ (c) $V_{34} = -2$ (d) $V_{24} = 0$

(e) $V_{13} = +5$ (f) $V_{14} = +3$ (g) $V_{32} = -2$

1.11   A resistor and battery are connected to the ground terminals of the scope as shown below. What is the current flowing through the resistor (if any) ? _____ What is the voltage be-   (a)
tween terminals 2 and 4 ? _____   (b)

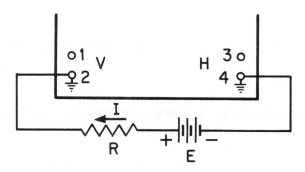

The spot will move across the screen once (at a uniform rate) in T seconds and then return to its starting point.

2.11   The rate at which the spot sweeps across the screen is variable by means of the SEC/DIV switch. The screen is 10 divisions wide. If the sweep rate is set at 50 msec/div, the time required for the spot to cross the screen once will be _____.   (a)

If the sweep rate is set to R sec/div, the time required for the spot to travel x div is

t = _____   (b)

No, because the impedance of the blocking capacitor is negligible over the frequency range where the AC input is normally used.

3.11   The vertical amplifier is set to DC and S_v = 1 volt/div, and the input is connected as shown. The deflection of the spot is

_____ divisions. If the scope were an ideal measuring   (a)

instrument, the deflection would be _____ divisions.   (b)

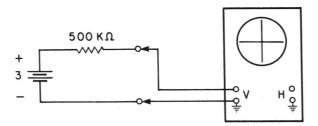

(a) + + − −   (c) $0 < \theta < 90°$ (d) $-180 < \theta < -90°$

(b)

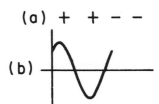 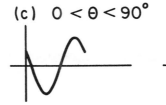 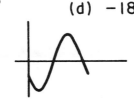 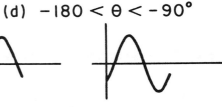

4.11   The scope is set to internal trigger with trigger slope set to + and trigger level set so that the sweep triggers when the input voltage goes through 0. The applied vertical input voltage is $v = 10 \sin(\omega t + \theta)$. Sketch the waveform which will be observed on the screen, paying particular attention to the point at which the waveform starts.

(a)

Does the picture observed depend on the value of θ? _____   (b)

Explain why or why not _____   (c)

Can we determine the value of θ by observing the waveform and using

the same waveform to trigger the sweep? _____   (d)

24

(a) E/R (since the ground terminals are connected together internally)

(b) 0

1.12 If sources are connected to the scope terminals as shown below, indicate a possible location for the spot on the diagram.

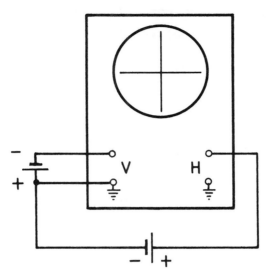

(a) 500 ms (1/2 sec)   (b) $t = (R \text{ sec/div}) \cdot (x \text{ div})$

2.12 If the sweep rate is set so that the spot will take 3 seconds to cross the screen and the sweep generator is triggered <u>once</u> at t = 1, sketch the sweep waveform which will be generated.

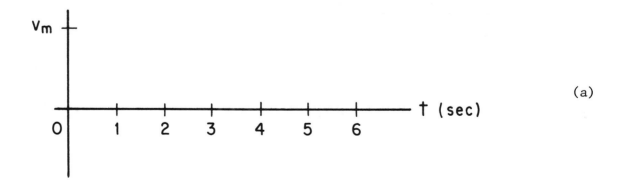

(a)

For the above example, the sweep rate is _____ sec/div.   (b)

(a)   2 (since the input voltage is $\dfrac{1 \text{ meg}}{.5 \text{ meg} + 1 \text{ meg}}$ 3 = 2)

(b)   3

Use of the Scope Probe to Reduce Loading

3.12   The deflection is less than the correct value because the scope loads down the circuit. How could the loading effect of the scope be decreased?

_____

If your answer to 4.11 (b) is YES, continue with this frame. Otherwise, turn to frame 4.13.

4.12   Assume the same input voltage is applied to the sweep trigger and the vertical amplifier. Since the trigger level is set to 0, the value of the input voltage when the sweep triggers will be _____.
Therefore, the waveform will start when the input voltage is _____.
Now turn back to 4.11 and try again.

anywhere in the 4th quadrant

1.13   The following voltages are desired at the scope input terminals:

  vertical input voltage (v_v) = +3 volts

  horizontal input voltage (v_h) = -3 volts

The arrangement shown below is proposed. Will it provide the required voltages? _____ (a)

If not, explain. _____ (b)

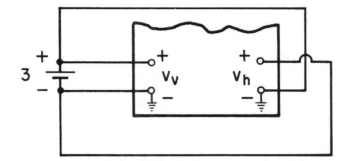

(a)

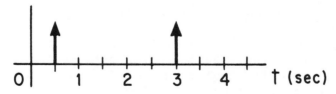

(b) $\dfrac{3 \text{ sec}}{10 \text{ div}}$ = .3 sec/div

2.13   If the sweep rate is set at .1 sec/div and the trigger pulse input to the sweep generator is as follows

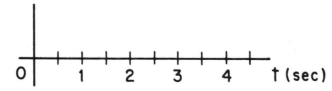

Sketch the sweep waveform which will be generated.

If you need a hint, look at Fig. 5-4 on p. 235.

by increasing the input impedance

3.13 A probe* is available to increase the effective input impedance of the scope. The probe also attenuates (reduces) the input signal so that the deflection is less for a given input and sensitivity setting.

If a probe which has a DC impedance of 9 megohms is connected to the scope input as shown, the effective DC input impedance of the scope (including the probe) is _____. (a)

If a DC voltage of V_p volts is applied to the input terminals of the probe, the voltage at the scope input terminals would be _____. (b)
This probe attenuates a DC input signal by a factor of _____. (c)

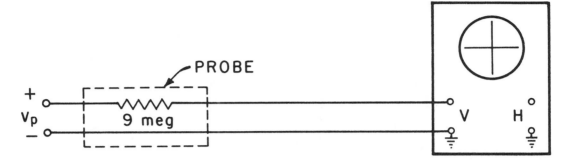

*In this and subsequent frames, the word probe will refer to a 10X attenuating probe.

Answers to 4.11:

(a)

(b) NO

(c) the waveform starts when v = 0 and has a positive slope since that is when the sweep triggers

(d) NO

Phase Shift Measurement by the Dual Trace Method

4.13 In the following frames, you will learn how to measure the phase angle between two sinusoidal voltages using the dual trace method. In this method, you will display one voltage using the CH1 vertical input and the other voltage using the CH2 vertical input. If the time axis is calibrated in degrees, you can then read the phase angle directly from the scope screen.

(a) If your answer is YES, continue with this frame. Otherwise, turn to
      frame 1.15.

1.14 Yes would be the correct answer __if__ the ground terminals were independent.

The voltage between the ground terminals on the scope is always _____ (a)
because _____ (b)

Now turn to frame 1.13 and try again.

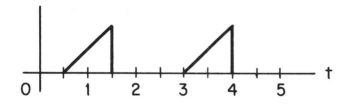

(Note that the duration
of the sweep waveform
is .1 sec/div x 10 div=
1 sec.)

2.14 If a periodic waveform is used as a triggering signal, the sweep is
triggered periodically. For the following sweep waveform the sweep
is triggered once every _____. The spot is at the left edge   (a)
of the screen (x = 0) between t = _____ and _____,
t = _____ and _____, etc. The sweep rate is   (b)
_____ sec/div. (Assume that v_m is just sufficient to deflect   (c)
the spot the full screen width of 10 divisions.)

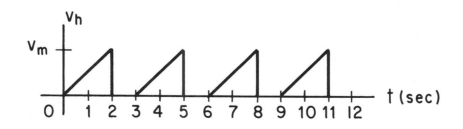

(a) 10 megohms (9 megohms from the probe and 1 megohm from the scope)

(b) $V_p/10$ (c) 10

3.14 The probe is now connected to the same source as was used in frame 3.11
(see below). To compensate for the probe attenuation, we will use
$S_v = 0.1$ volt/div instead of 1 volt/div. The voltage at the probe

input terminals (V_p) is _____. The voltage at the scope input   (a)

terminals is _____. Since $S_v = 0.1$ volt/div, the deflection   (b)

of the spot is _____. Compare this with frame 3.11 and   (c)

note that the loading effect is reduced when the probe is used.

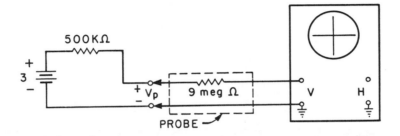

4.14 (1) The scope vertical mode is set to CH1 and the trigger source is set
to internal (from CH1). $v_1 = \sin \omega t$ is applied to the CH1 vertical
input and the trigger slope and level controls are adjusted to obtain
the picture shown in Fig. 4-14(a).

(2) With the scope controls still set as in (1) above, the vertical mode is
changed to CH2 (with internal trigger from CH2). The CH2 vertical
input is $v_2 = \sin(\omega t + 45°)$. The resulting trace will be as shown in

Fig. 4-14 (a/b/c) _____.   (a)

(3) With the CH1 and CH2 inputs still $v_1 = \sin \omega t$ and $v_2 = \sin(\omega t + 45°)$,
the vertical mode is set to DUAL TRACE (internal trigger from CH1).
The resulting <u>CH2</u> trace would be as shown in Fig. 4-14 _____.

Explain. _____   (b)

_____

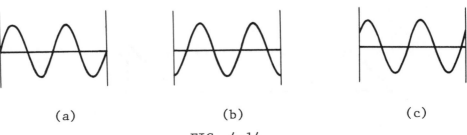

(a)   (b)   (c)

FIG. 4-14

Answers to 1.13:

(a) NO is the correct answer.

(b) The battery is shorted out by the internal connection between the ground terminals.

Deflection of the Spot by DC Inputs

1.15 The scope screen has grid lines superimposed on it so that the amount of deflection is easy to observe. In the diagram below, the vertical deflection of the spot is

_____ divisions   (a)

and the horizontal deflec-

tion of the spot is

_____ divisions   (b)

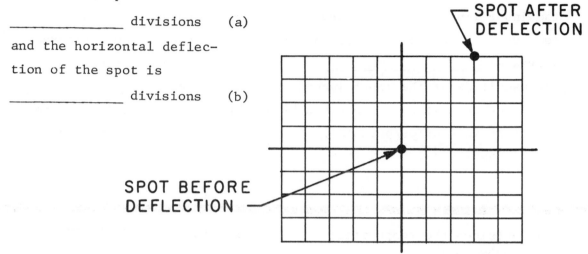

SPOT AFTER DEFLECTION

SPOT BEFORE DEFLECTION

(a) 3 sec   (b) 2 and 3, 5 and 6, etc.   (c) .2 sec/div

2.15 Once the sweep generator has been triggered, it cannot be retriggered until the sweep is completed. If a trigger pulse occurs during the middle of a sweep waveform, this trigger pulse will have no effect. In the following diagram, what will be the effect of a trigger pulse which occurs at t = t_2? _____ At t = t_3?   (a)

_____   At t = t_4? _____   (b,c)

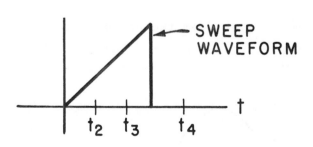

SWEEP WAVEFORM

t_2 t_3 t_4 t

(a) $\quad 3 \dfrac{10M}{10M + .5M} = 2.86$ volts $\qquad$ (b) $\dfrac{2.86}{10} = .286$ volts

(c) $\quad \dfrac{.286}{.1} = 2.86$ div. (compared with 2 without the probe)

3.15 We have seen how a probe could be used to raise the input impedance of
the scope and reduce the loading for DC (or low frequency) signals.
We will now turn our attention to high frequencies. At high frequencies
the input impedance of the scope looks like a capacitor. In order
to reduce the capacitive loading due to the scope, we must (increase/
decrease) _____ the input impedance. Since $\qquad$ (a)
the magnitude of the input impedance is $1/\omega C$, we must _____ $\qquad$ (b)
the input capacitance. This can be accomplished by placing another
capacitor in (series/parallel) _____ with the scope $\qquad$ (c)
input capacitance.

(a) a (since the scope still triggers when the level is 0 and the
     slope is +)

(b) c. The scope now triggers at t = 0 so the waveform begins at
     $v_2 = \sin 45^\circ$.

4.15 We have two voltage sources $v_1 = A \sin \omega t$ and $v_2 = B \sin (\omega t + \theta)$.
Before we display v_2 on the screen, we must determine the location of
the origin (t = 0) using the CH1 input. To locate the origin, we will
first display v_1 and adjust the triggering controls to obtain the picture
shown at the bottom of the page. Label the horizontal axis with the appropriate number of degrees at each division. $\qquad$ (a)

Next we will connect v_2 to the CH2 input and set the scope for dual trace
operation. In order to maintain the origin (t = 0) at the same point on
the screen, we must be certain both sweeps will trigger from (v_1/v_2)
_____. $\qquad$ (b)
For the sweep to trigger from the appropriate vertical input signal, the

triggering SOURCE should be set to (INTernal, EXTernal) _____. $\qquad$ (c)
On the figure below sketch the CH2 trace, v_2, when $\theta = -120^\circ$.

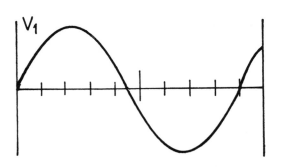

$\qquad\qquad$ (d)

(a) 4   (b) 3

1.16   The vertical deflection, y, is proportional to the
vertical input voltage, v_v. The amount of deflection obtained for a given v_v depends on the vertical
sensitivity. The sensitivity is defined as the
number of volts required to deflect the spot one
division. If the vertical sensitivity is S_v volts/
division and vertical input is v_v volts, the vertical deflection of the spot is y = _____
divisions. Make sure your answer is dimensionally
correct.

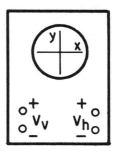

(a) No effect   (b) No effect   (c) the sweep will be triggered
                                             again

2.16   If the sweep rate is set at .5 sec/div and the trigger pulse input to the
sweep generator is as follows:

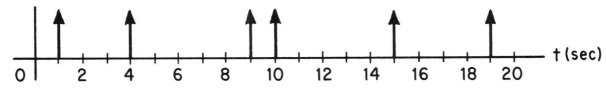

plot the horizontal (x) deflection of the spot as a function of time:

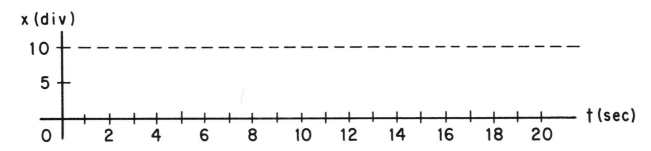

33

(a) increase (since the loading effect is less when the input impedance is high)

(b) decrease

(c) series (series capacitors combine like parallel resistors)

3.16   Let us consider how a voltage divides across two capacitors in series. (Assume that the capacitors are initially uncharged.)

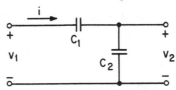

If a current i flows in the series combination:

$$v_2 = \frac{1}{C_2} \int_0^t i(t)\,dt$$

$$v_1 = \underline{\hspace{5cm}} + \underline{\hspace{4cm}}$$ (a)

$$\frac{v_2}{v_1} = \frac{\underline{\hspace{3cm}}}{\underline{\hspace{1cm}} + \underline{\hspace{1cm}}} = \frac{\underline{\hspace{3cm}}}{\underline{\hspace{1cm}} + \underline{\hspace{1cm}}} = \frac{\underline{\hspace{3cm}}}{\underline{\hspace{1cm}} + \underline{\hspace{1cm}}}$$ (b)

Does the voltage ratio depend on the frequency of the waveform of i? $\underline{\hspace{3cm}}$ (c)

(b) v_1

(c) INTernal

(d)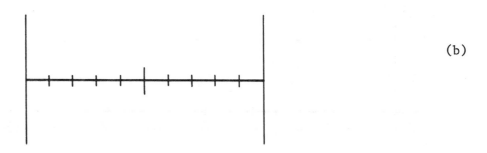

4.16   In the previous frame, each division corresponds to $\underline{\hspace{4cm}}$ (a) degrees. To improve the accuracy with which we can determine the phase angle, we will change the time scale so that each division corresponds to 20°. After this change, sketch the waveform we should see when the vertical input is $v_1 = A \sin \omega t$.

   (b)

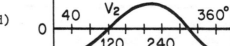

1.17   If the vertical gain is set at 5 volts/div, an input voltage of 10

volts will deflect the spot <u>upward</u> _____ divisions.   (a)

To deflect the spot 3 divisions <u>downward</u>, an input voltage of

_____ volts is required.   (b)

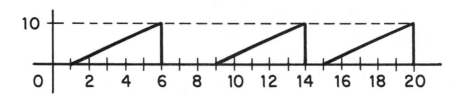

THIS IS A GOOD PLACE TO TAKE A BREAK

<u>Trigger Slope and Level Controls</u>

2.17   We have just studied the operation of the sweep generator. The output of the sweep generator is used to drive the horizontal amplifier
when we display a waveform on the screen as a function of time. When
a trigger pulse occurs at the input of the sweep generator, a sweep
waveform is generated which sweeps the spot across the screen once.
When the next trigger pulse occurs, the sweep will be triggered again
provided that the previous sweep has been completed.

As shown in Fig. 5-4 (p. 235) the input pulses to the sweep generator
come from the _____. If there   (a)
is no triggering signal present at the input of the trigger pulse
generator no output pulse will occur. In order to trigger the
sweep generator, a _____ must be   (b)
present at the input of the trigger pulse generator.

(a) $v_1 = 1/C_1 \int_0^t i(t)dt + 1/C_2 \int_0^t i(t)dt$

(b) $\dfrac{v_2}{v_1} = \dfrac{1/C_2 \int_0^t i(t)dt}{1/C_1 \int_0^t i(t)dt + 1/C_2 \int_0^t i(t)dt} = \dfrac{1/C_2}{1/C_1 + 1/C_2} = \dfrac{C_1}{C_1 + C_2}$

(c) No (The voltage ratio is independent of frequency and waveshape since i cancels out.)

3.17 Previously we used a resistive voltage divider to raise the input impedance of the scope:

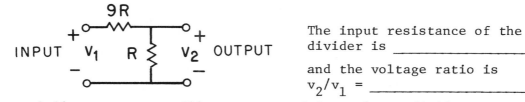

The input resistance of the divider is _____ (a)

and the voltage ratio is $v_2/v_1 =$ _____ (b)

In a similar manner we will use a capacitive voltage divider to reduce the input capacitance. How does the frequency and waveshape of v_1 effect the voltage ratio of the divider? _____ (c)

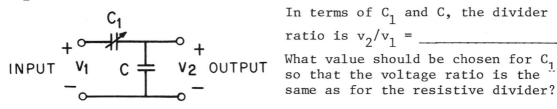

In terms of C_1 and C, the divider ratio is $v_2/v_1 =$ _____ (d)

What value should be chosen for C_1 so that the voltage ratio is the same as for the resistive divider?

_____ (e)

With this value of C_1 the input capacitance is _____ (f)
(Remember that series capacitors combine like parallel resistors.)

(a) 40° (b)

4.17 The two steps used in measurement of phase shift by the dual trace method are:

(1) Calibrate the time axis (in degrees) using v_1 (CH1).

(2) Observe v_2 (CH2) in the dual trace mode using the same time axis.

Once the time origin is set in step 1, the triggering circuit must not be changed in any way. Therefore, in both steps 1 and 2 the triggering signal must be (v_1/v_2) _____ and the triggering (a)
SOURCE set to _____ . (b)

(a) $\dfrac{10 \text{ volts}}{5 \text{ volts/div}} = 2$ div   (b) -3 div $\times$ 5 volts/div $= -15$ volts

---

1.18   The horizontal deflection, x, is proportional to the horizontal input voltage, v_h. The gain of the horizontal amplifier determines the horizontal sensitivity, S_h, measured in volts/division. The horizontal deflection (in divisions) is given by the following expression:

x = _____

(a)   trigger pulse generator
(b)   triggering signal

2.18   The point on the triggering signal at which the sweep is triggered depends on the setting of two switches, TRIGGER SLOPE AND TRIGGER LEVEL. The trigger slope determines whether the sweep is triggered on the positive or negative slope of the waveform. If the TRIGGER SLOPE is negative and the triggering signal is as shown, the sweep could be triggered at some time in region _____ or _____ but not in _____ or _____.

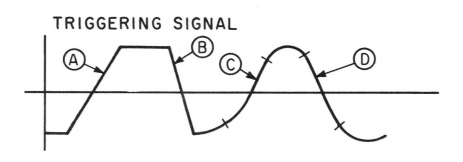

TRIGGERING SIGNAL

(a)   10R   (b)   1/10   (c)   No effects   (d) $C_1/(C_1 + C)$

(e) $\dfrac{C_1}{C_1 + C} = \dfrac{1}{10}$ so $C_1 = \dfrac{C}{9}$

(f) $\dfrac{C \left(\frac{C}{9} \right)}{C + \frac{C}{9}} = \dfrac{C}{10}$

3.18   Now consider using both voltage dividers simultaneously:

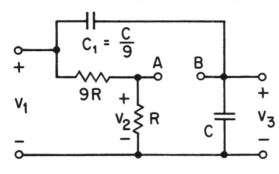

With C_1 adjusted so that $v_2 = v_3$, how much current will flow in a wire connected from A to B? _____   (a)

How will connecting a wire from A to B affect the behavior of the network?

_____   (b)

What is the voltage ratio v_2/v_1 with a wire from A to B?

_____   (c)

As the frequency of the input signal is increased, this ratio will (increase/remain constant/decrease) _____   (d)

(a) v_1

(b)   INTernal (CH1)

4.18   For the following network, $v_1 = A \sin \omega t$ and $v_2 = B \sin (\omega t + \theta)$. θ is to be measured by the dual trace method.

Which voltage should be observed in step 1? _____   (a)

Which voltage should be observed in step 2? _____   (b)

For both steps, the triggering signal should be _____.   (c)

Why must the triggering controls be set exactly the same for both

steps? _____   (d)

38

Check your own answer to make sure that it is dimensionally correct.

1.19 | If the horizontal sensitivity is 0.5 volts/div, 2 volts will deflect
the spot _____ divisions to the right. How much, and in what   (a)
direction will -1.5 volts deflect the spot? _____   (b)

B or D, but not in A or C.

2.19 | The TRIGGER LEVEL control is variable from - to + and determines the
value (sign and magnitude) of the triggering voltage at which the
sweep is triggered. If the TRIGGER LEVEL control is set at 0, the
sweep is triggered when the waveform passes through 0. The TRIGGER
LEVEL and SLOPE controls are adjusted so that the sweep is triggered
at point B on the waveform shown below. If we want the sweep to be
triggered at point A instead of B, we should set the TRIGGER

_____ control to _____.   (a,b)

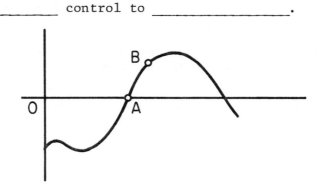

(a)   None   (b)   No effect   (c) $\frac{1}{10}$ (d)   Remain Constant

---

3.19   For the same network with no connection between A and B, draw an
        equivalent <u>input</u> circuit consisting of one resistor and one capacitor.

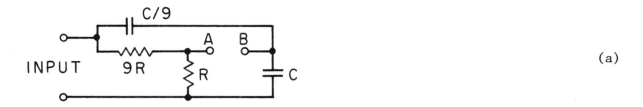

                                                                                        (a)

Now draw the equivalent circuit when A and B are connected (<u>one</u> R and
<u>one</u> C).

                                                                                        (b)

(a) v_1 (b) v_2 (c) v_1
(d)   so the time origin will be the same

4.19   Indicate the proper connections to the scope for calibrating the time
        axis (step 1). Now add the connection(s) necessary to view v_2 on the
        same time axis (step 2).

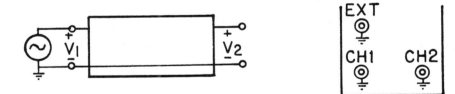

    Note: The line drawn through the lower part of the network indicates
            that there is a common connection between v_1- and v_2-

1.20 The grid lines on the scope screen are reproduced below.

Assuming that the spot is initially centered on the screen when no voltages are applied, plot the location of the spot for each of the following conditions:

(a) S_v = 5 v/div, v_v = -15 v;

   S_h = .1 v/div, v_h = .2 v

(b) S_v = 10 mv/div, v_v = 30 mv;

   S_h = 2 v/div, v_h = -4 v

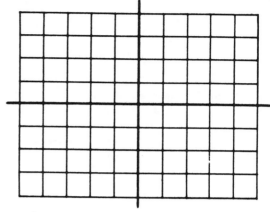

2.20 A sine wave is used as a triggering signal. The resulting trigger pulses and sweep waveform are shown below:

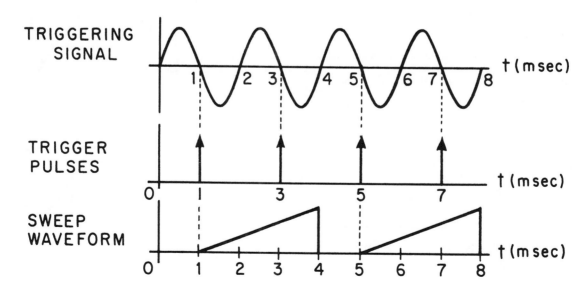

The trigger slope is set to _____. Trigger level is set to _____.(a,b)

Sweep time/div is set to _____. The sweep generator is not   (c)

triggered by pulses at t = _____ or t = _____ because   (d)

_____. (e)

41

(a)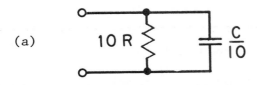

(b) the same (since connecting A to B doesn't effect circuit operation)

3.20 Now let's get back to the probe and the scope. If the probe contained only a resistor as shown below, as the input frequency was increased, the ratio V_v/V_p would _____ (a)

If we want V_v/V_p to be independent of frequency indicate on the diagram what must be added to the probe. If in doubt, go back and restudy frame 3.18. (b)

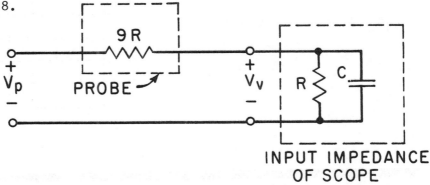

INPUT IMPEDANCE OF SCOPE

---

v_1+ to CH1 input, v_2+ to CH2 input, either v_1- or v_2- to scope ground

---

4.20 The sweep is calibrated with $v_1 = A \sin \omega t$ and the controls are adjusted to obtain Fig. 4-20x. Then $v_2 = B \sin(\omega t + \theta)$ is observed with the same time axis.

If $v_2(0) > 0$, then θ is a (+,-) _____ (a) angle between 0 and 180°.

If $v_2(0) < 0$, then θ is a _____ angle (b) between 0 and 180°.

For Fig. 4-20y give an equation for θ in terms of θ_1. $\theta = $ _____ (c)

For Fig. 4-20z give an equation for θ in terms of θ_2. $\theta = $ _____ (d)

Before you turn the page, check your answers to (c) and (d) by considering the limiting cases as θ_1 and θ_2 approach zero.

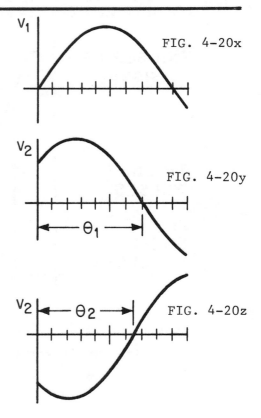

FIG. 4-20x

FIG. 4-20y

FIG. 4-20z

(a) plot at x = 2, y = -3

(b) plot at x = -2, y = 3

1.21 The vertical and horizontal inputs to the scope are not completely independent because their ground terminals are _____

_____.   (a)

When external voltages are applied to both the vertical and horizontal
scope inputs, this is frequently referred to as the X-Y mode of operation.
(Later we will study the sweep mode in which an internally generated voltage is used to deflect the spot in the X direction). In the previous
frame, the scope is used in the _____ mode of operation.   (b)

(a) - (negative)   (b) 0   (c) 0.3 msec   (d) 3 msec or 7 msec

(e) the previous sweep waveform has not gone to completion

2.21 A trigger pulse will occur at the output of the trigger pulse generator (Fig. 5-4) provided that two conditions are met:

(a) the _____ of the triggering signal matches the

setting of the _____ control and   (a)

(b) the _____ of the triggering signal matches the

setting of the _____ control.   (b)

The trigger pulse will then trigger the sweep generator provided

that _____   (c)

(a) decrease   (b)

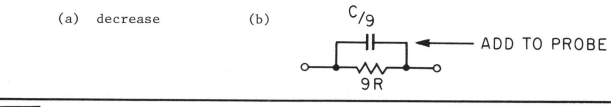

9R — ADD TO PROBE

3.21

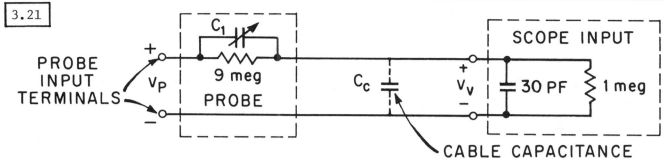

CABLE CAPACITANCE

Since the coaxial cable which connects the probe to the scope has shunt capacitance, this capacitance must be added in parallel with the scope input capacitance as shown. If C_1 is properly adjusted so that the divider ratio V_v/V_p is constant, <u>sketch</u> the equivalent circuit seen at the probe input terminals (V_p). Assume C_c = 70 pf.

← (a)

For DC, the input signal is attenuated (reduced) by a factor of _____.
At all frequencies the input signal is attenuated by a factor of _____.

(b)

(c)

TURN TO FRAME 4.21

44

(a)   connected together (connected to the case and chassis)   (b)   X-Y

1.22 | In this and the following frames, assume $S_v = S_h = 1$ volt/division. Also
assume that the spot is centered when no inputs are applied.

A student has connected inputs
to the scope as shown.

The vertical deflection will be

_____ div.   (a)

The horizontal deflection will

be _____ div.   (b)

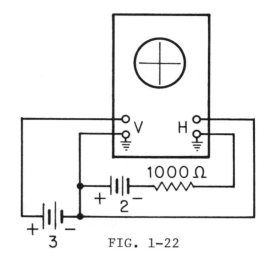

FIG. 1-22

(a)   slope, (TRIGGER) SLOPE   (c)   the previous sweep waveform has gone
(b)   level, (TRIGGER) LEVEL   to completion

2.22 | In the case illustrated below, the TRIGGER SLOPE control is set to + and
the LEVEL control is set to 0. Sketch the trigger pulses and the resulting
sweep waveform if the duration of the sweep is T_1 seconds.

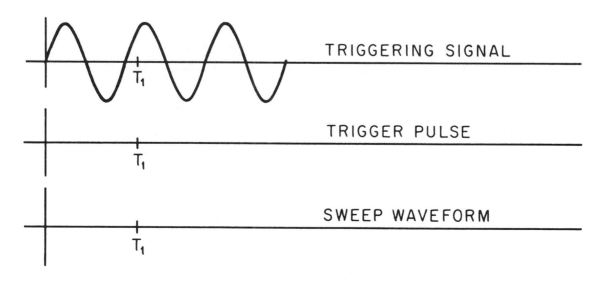

TRIGGERING SIGNAL

TRIGGER PULSE

SWEEP WAVEFORM

(a)

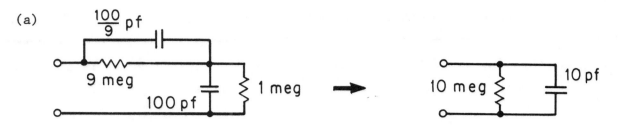

(b) V_v/V_p = 1 meg/(9 meg + 1 meg) = 1/10, so reduction is by a factor of 10

(c) 10 (since attenuation is independent of frequency)

3.22 The probe is used to measure the amplitude of a sine-wave source as shown below.

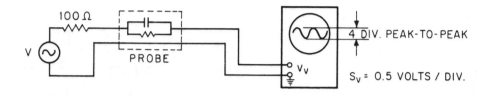

The peak-to-peak amplitude of V is _____ volts.

(a) +   (b) −   (c) $\theta = 180 - \theta_1$ (d) $\theta = -\theta_2$

4.21 With the scope set as in frame 4.16 (20°/division), the dual trace display is observed for various values of θ. Give the value of θ in each case:

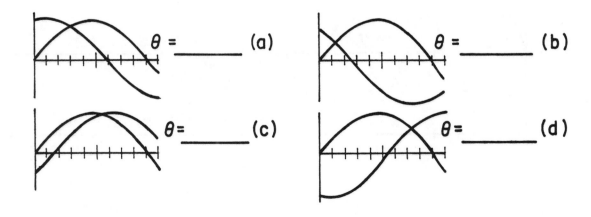

$\theta =$ _____ (a)

$\theta =$ _____ (b)

$\theta =$ _____ (c)

$\theta =$ _____ (d)

(a)   3 div

(b)   If your answer is 1, 2, 3, -1, -2, or -3 continue with this frame
      (1.23).   Otherwise, turn to frame 1.24.

1.23   Reexamine Fig. 1-22.   The voltage which would be measured between the
two ground terminals is not 1, 2, 3, -1, -2, or -3 volts.   What

would it be? _____   (a)

Note how the H input is connected.   Now what is the horizontal deflection. _____   (b)

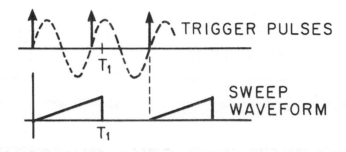

2.23   The waveform shown in Fig. 2-23a is used as the vertical input (v_v) and as
the triggering signal.   If the control settings are TRIGGER LEVEL = 0,
TRIGGER SLOPE = +, sweep rate = .5 sec/div, sketch (on Fig. 2-23b) the
sweep waveform which is generated.

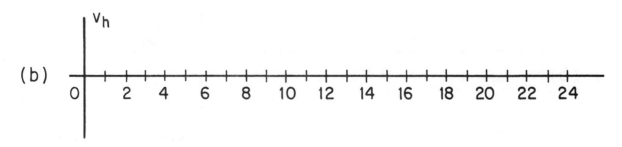

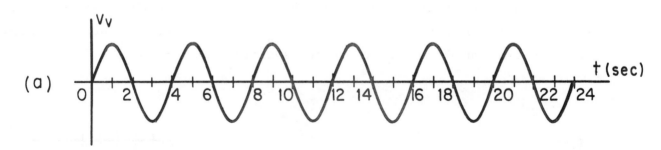

FIG. 2-23

47

20 volts (since V_v = 2 volts peak-to-peak and the probe attenuates the signal by a factor of 10).

3.23 When making accurate measurements with the scope, the probe should be used if the impedance of the source being measured is not small compared with the scope input impedance. In which of these cases should the probe be used? _____

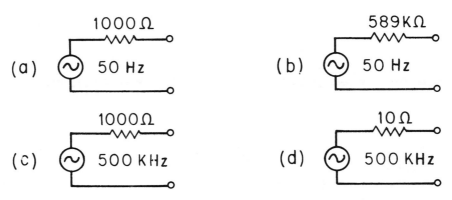

NOTE: The input impedance of the scope at 500 kHz is approximately 10 kilohms.

(a) 60°   (b) 130°   (c) −30°   (d) −110°

4.22 Given v_1 = A sin ωt, v_2 = B sin(ωt + θ), −90° < θ < 0°

Explain (a) how you would calibrate the scope so that each division represents 10° and (b) how you would determine θ. State where v_1 and v_2 should be connected and sketch the expected waveforms for θ = −45°.

(a)

- -

(b)

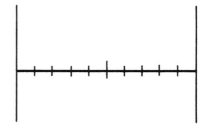

Answer to 1.22(b) and 1.23(b): The horizontal deflection is 0 (because the 2 volt source and series resistor is shorted out by the ground terminals).

This is a good place to take a break before you continue with the program.

Scope Traces Due to Time-Varying Inputs

1.24 │ Up to this point, we have learned how to determine the resulting deflection of the spot when constant voltages are applied to the input terminals. We are now going to investigate the patterns which are traced out on the scope screen when we apply time-varying voltages to the input terminals.

Unless specified otherwise, still assume that

$$S_v = S_h = 1 \text{ volt/div.}$$

Turn to frame 1.25.

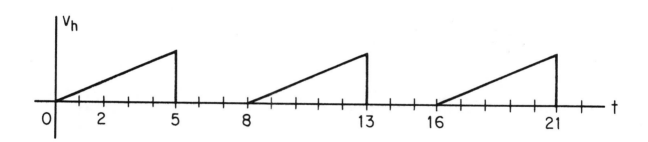

2.24 │ Now sketch the waveform which appears on the screen.

Assume that the spot starts at the left edge of the graticule when the sweep signal is 0, and assume that the peak value of the sweep voltage is sufficient to deflect the spot the full width of the screen.

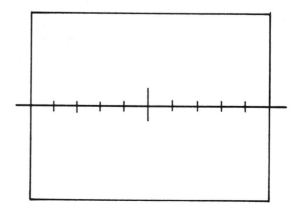

(b) and (c)

(Not (a) because 1000 << 1 megohm;

not (d) because   10 << 10 kilohms)

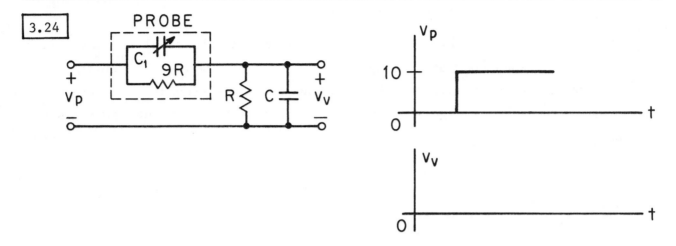

3.24   PROBE

Assuming that the probe capacitance is properly adjusted, sketch v_v if v_p is as shown.

(a) Connect v_1 to the CH1 vertical input. With the scope set on internal trigger, adjust for the picture at the left.

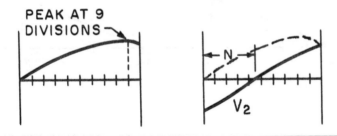

PEAK AT 9 DIVISIONS

V_2

(b) Connect v_2 to the CH2 vertical input. In the dual-trace mode, read the number of divisions as indicated.

$\theta = -(N \text{ div.} \times 10^{\circ}/\text{div.})$

Phase Measurement by the Triggered Sweep Method

4.23   The triggered sweep method for phase shift measurement is similar to the dual trace method, except that v_1 and v_2 are not viewed at the same time. Use of the triggered sweep method permits phase shift measurements to be made on single trace scopes, and it also permits use of the differential mode on the dual trace scopes.

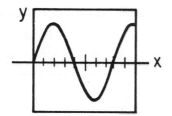

Note that 5 seconds of the waveform
(1-1/4 cycles) appear on the scope
screen.

2.25   In the preceding two frames, we had TRIGGER LEVEL = 0, TRIGGER SLOPE = +,
and SWEEP RATE = 0.5 sec/div. The TRIGGER SLOPE is now changed to -. The
vertical input and triggering signal are still the same.

Sketch the sweep waveform which is generated (on Fig. 2.25b).

Then sketch the waveform which appears on the screen (on Fig. 2.25c). Note
that the trace always starts at the left edge of the screen no matter what
time the sweep is triggered.

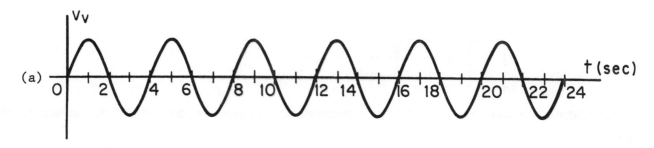

(a)

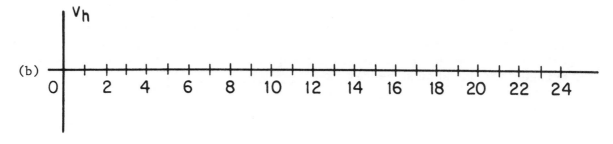

(b)

FIG. 2-25   (c)

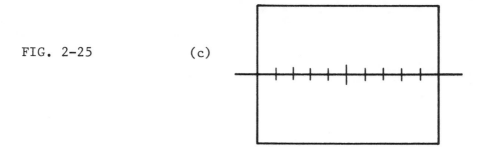

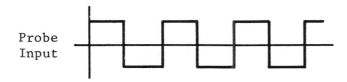

3.25   If the probe capacitance is not properly adjusted, the input voltage at the
scope terminals will be distorted. In which of the following case(s) is
the probe capacitance properly adjusted? _____

Probe
Input

Waveforms displayed on screen:

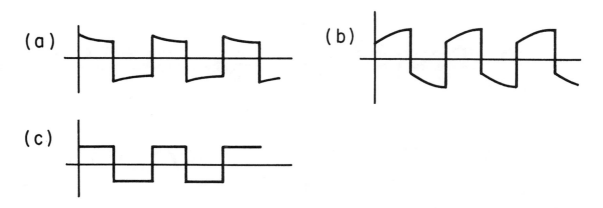

4.24 $v_2 = \sin(\omega t - 90^\circ)$ is applied to the CH1 vertical input of a scope and
$v_1 = \sin \omega t$ is applied to the EXTernal trigger input. If the trigger
slope is + and the trigger level is 0, sketch the waveform which will
appear on the screen if (a) internal trigger is used (b) external trigger
is used. If we want to observe the phase angle of v_2 relative to v_1, the
trigger source should be set to _____.

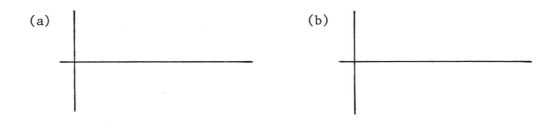

1.25 If the input voltage is time-varying, the spot will move across the screen. If $v_v = 0$ and v_h is as shown, indicate the position of the spot at $t = 0, 1, 2, 3,$ and 4 on the grid. Label the points $t = 0$, $t = 1$, etc.

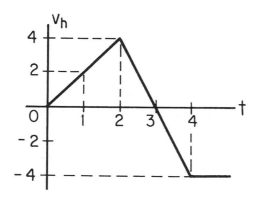 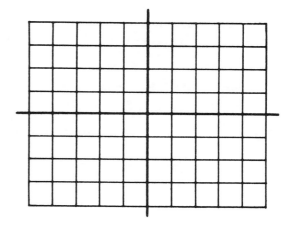

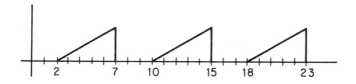 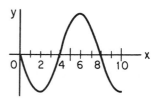

2.26 The waveform given below is used as a triggering signal. If the TRIGGER SLOPE is set to +, the sweep could trigger at points _____, _____, _____ depending on the setting of the level control.

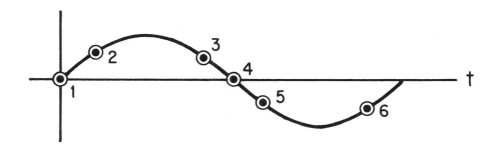

(c) only (in (a) and (b) the waveforms are distorted)

3.26 What is the effect of the probe on the scope input impedance?

_____ (a)

_____

When should the probe be used? _____ (b)

_____

The vertical sensitivity is set to S_v volts/div (without the probe).
When the probe is connected, if the vertical deflection is y divisions, the probe input voltage is _____. (c)
Given a square wave as a vertical input signal, how could you tell if
the probe capacitance was adjusted correctly? _____

_____ (d)

(a)

(b)

EXTernal

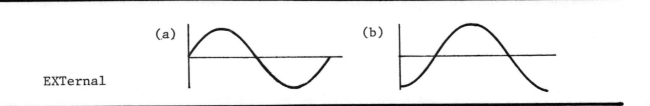

4.25

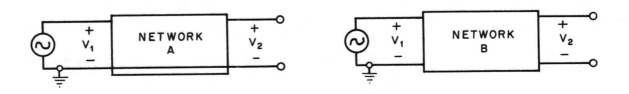

Network A has a common connection between v_1- and v_2-. Network B does not
have a direct connection between v_1- and v_2-. In both cases, the scope
ground is connected to the oscillator ground. To observe v_2 on the scope,
both the +INPUT (CH1) and −INPUT (CH2) on the differential amplifier must
be used for network _____ because _____

_____

Plot on the x-axis as follows:

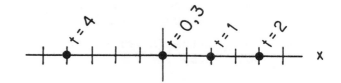

1.26 | With $v_h = 0$ and $v_v = 2 \sin (2\pi \cdot 3t)$, the spot will move up and down

between the limits _____ divisions and (a)

_____ divisions at the rate of (b)

_____ <u>cycles</u> per second. (c)

---

1   2   6

2.27 | In order to trigger at points 3, 4, or 5, the TRIGGER SLOPE must be

set to _____. (a)

If the TRIGGER LEVEL control is set in the positive range, triggering will occur when the triggering signal is _____. (b)
To cause triggering to occur during the negative half cycle of the

triggering signal, the TRIGGER _____ (c)
should be set to _____. (d)

(a) the input impedance is increased by a factor of 10

(b) if the impedance of the circuit being measured is not small compared with the scope input impedance

(c) 10 S_vy (the factor of 10 comes in because the probe divides the input voltage by 10)

(d) adjustment is correct if the waveform on the screen is an undistorted square wave

THIS IS A GOOD PLACE TO TAKE A BREAK

Other Measurement Errors

3.27 We have just studied how the probe can be used to reduce measurement errors which are due to loading. We will now consider several other sources of measurement errors.

There is a limit to how closely we can read the deflection from the screen, especially if we have to interpolate between divisions. It is usually difficult to read the deflection closer than the nearest 1/10 of a major division. Hence there may be a reading error of ± 1/20 division (or more if we aren't careful). Assuming a reading error of ± 1/20 divisions, what is the per cent reading error if the deflection is 4 divisions? _____ (a)

If the deflection is 1 division? _____ (b)

To minimize the per cent reading error, we should _____ (c)

_____.

B because grounding v_2– would short out part of the network

TURN TO FRAME 4.26

(a) +2   (b) −2

(c) 3 (Remember that $f = \omega/2\pi$ where f is the frequency in cycles/sec (Hz);
ω is in radians/sec.)

1.27 If the frequency of the sine wave, $v_v = 2 \sin \omega t$,
is increased so that we cannot follow the motion
of the spot with our eye, sketch what we will see
on the grid below. (v_h still equals 0.)

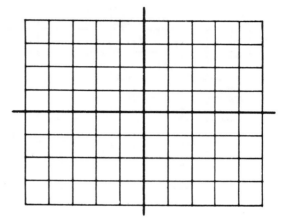

(a) − (minus)

(b) positive   (c) level   (d) (−) the negative range

2.28 For the given waveform, fill in the points at which the sweep will trigger
for the indicated control settings:

	LEVEL −	LEVEL 0	LEVEL +
SLOPE +			
SLOPE −			

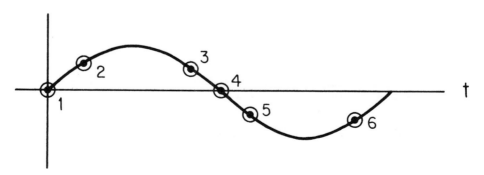

FIG. 2-28

(a) $\dfrac{1/20}{4} \times 100\% = 1.25\%$ (b) $\dfrac{1/20}{1} \times 100\% = 5\%$

(c) use as large a deflection as possible

3.28 With $S_v = .2$ volts/div, the following waveform is displayed on the screen.

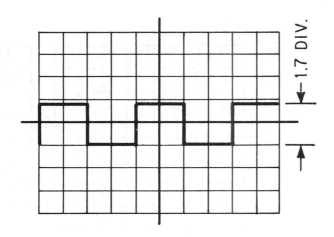

↕ 1.7 DIV.

Assuming a reading error of $\pm 1/20$ division, the peak-to-peak <u>voltage</u> is

_____ $\pm$ _____ %   (a)

To minimize per cent reading error, change S_v^* to _____ (b)

Sketch the resulting trace in the proper position on the screen. (c)

With this setting, the error is _____ %   (d)

*Available values are 1, .5, .2, .1, .05, .02, etc. volts/div.

4.26 When the differential mode of operation is used, both the CH1 and CH2 inputs are required to display v_2, so the dual trace method of phase measurement cannot be used. The triggered sweep method uses the same two steps as the dual trace method:

(1) calibrate the time axis using v_1

(2) display v_2 using the same time axis

Step 2 requires using v_1 as a triggering signal when displaying v_2. On the diagram below, show the required connections to the scope to display v_2 as a function of time. (a)

Then show the connection which must be added if we want to trigger the scope with v_1. (b)

In this case, the trigger SOURCE switch should be set to _____ (c)

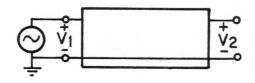
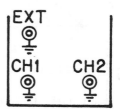

a vertical line on the y-axis between +2 and -2

1.28 With $v_v = 0$ and $v_h = 3 \sin \omega t$, the spot will move between the limits

_____ and _____ on the _____ axis.   (a,b,c)

Sketch the appearance of the scope trace at high frequencies on the

grid below.

(d)

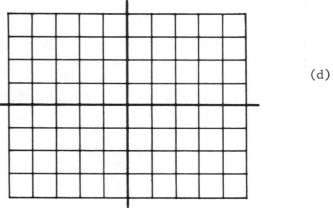

---

6 1 2   (Note that the functions of the SLOPE and LEVEL controls
5 4 3   are independent.)

2.29 The waveform of Fig. 2-28 is used as the vertical input and the controls
are set so triggering occurs at point 5.   SEC/DIV is set so that exactly
one cycle of the waveform appears on the screen.   Sketch the waveform as
it appears on the screen.

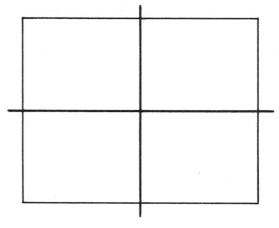

(a)   .34 ± 2.9% (1.7 div x .2 volts/div = .34; $\frac{\pm.05 \text{ div}}{1.7 \text{ div}}$ x 100% = 2.9%)

(b)   .05 volts/div

(c)   square wave between +2.8 and −4 divisions (6.8 div. peak-to-peak)

(d)   .74%

3.29

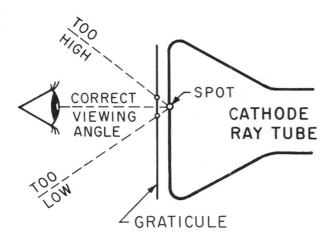

In some scopes, the graticule is a short distance in front of the screen. If the spot is viewed from too high or too low, it will appear in the wrong position with respect to the graticule lines as is shown in the diagram. This is known as parallax error. To avoid parallax error, one should view the scope

_____ (a)

_____

Most modern scopes have internal graticules, which means the graticule lines are etched right on the surface of the screen. This completely eliminates _____ _____ (b)

(a) v_2+ to +INPUT(CH1), v_2− to −INPUT(CH2), (scope in differential ampli-fier mode), function generator ground to scope ground

(b)   add v_1+ to EXTernal trigger input

(c)   EXTernal

4.27   The same time origin must be used for observing v_2 (step 2) as for calibrating the time axis (step 1); therefore the triggering circuit must not be changed in any way between the two steps. In both steps

1 and 2, v_1 must be connected to _____ and the trigger   (a)

source switch must be set to _____. In step 1, v_1 should also be   (b)

connected to the _____ input.   (c)

(a) +3   (b) -3   (c) x (horizontal)

(d) a horizontal line on the x-axis between +3 and -3

1.29 Plot the location of the spot when (1) $v_v = v_h = -2$, (2) $v_v = v_h = +1$,

(3) $v_v = v_h = +3$

If v_v and v_h are both functions of time with $v_v(t) = v_h(t)$, then at every instant of time the spot will lie on a _____ (a)
through the origin at an angle of _____. (b)

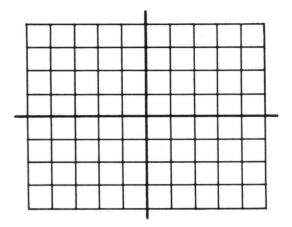

2.30 The level control is continuously variable from - through +. If the
level control is set more positive or more negative than the available triggering signal, the sweep will not be triggered. If the
level is set in the + range and no triggering is observed, either the
trigger level should be (increased/decreased) _____ (a)
or the input signal should be _____ in magnitude. (b)

(a)   directly in front of the screen

(b)   parallax error

3.30   In addition to the errors in reading the scope, there are errors which
are inherent in the scope itself. Even if the scope is properly
calibrated, there may be a calibration error of as much as ± 3%*. For
example if VOLTS/DIV is set to 1, the actual volts/division may be
anywhere between .97 and 1.03.

Assuming a calibration error of ± 3% in addition to a reading error
of ± 1/20 division, if S_v = .5 volt/div and the deflection is 5
divisions, the input voltage has a nominal value of _____, (a)
but it may actually lie anywhere in the range _____ to
_____.   (b)

_____

*5% for some scopes.

(a)   the EXTernal trigger input   (b)   EXTernal

(c)   CH1 vertical input

4.28   In the network shown below, v_1 and v_2 do <u>not</u> have a common ground.
Show the required connections to the scope for steps 1 and 2.

Step 1:

 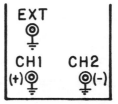

Step 2:

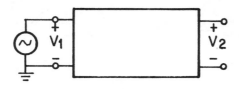 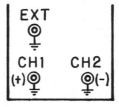

62

(a) line   (b) 45°

2.30 Sketch the appearance of the trace when

$$v_v(t) = v_h(t) = 4 \cos \omega t$$

Hint: Plot limiting values first.

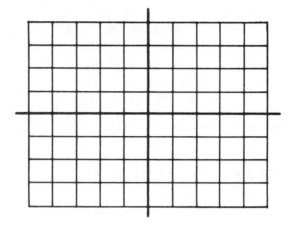

(a) decreased   (b) increased

THIS IS A GOOD PLACE TO TAKE A BREAK

Trigger Source

2.31 922 ONLY: Study Fig. 5-3(a), p. 234.

CAL X-Y ONLY: Study Fig. 5-3(b), p. 234.

ALL: Note that the triggering signal may come from 3 different sources.
If you want the triggering signal to have the same waveshape as the verti-
cal input, the TRIGGER SOURCE switch should be set to _____. (a)
If you want to trigger the sweep from a different source than the vertical

input or the AC line, the TRIGGER SOURCE should be set to _____

or _____.   (b)
The only difference between the EXT and EXT/10 positions is that in EXT/10

the external trigger input is divided by _____ before it is (c)
sent to the time base.

Is there a triggering signal applied to the TIME BASE when the TRIGGERING

SOURCE is set to X-Y? _____   (d)

(a)   2.5 volts   (b)   2.4 to 2.6 (2.5 ± 3% ± 1%)

3.31 Errors may occur if we try to use the scope with input signals which have too high a frequency.

Suppose that we use the scope to observe a cosine wave, A cos (2πft). The vertical sensitivity is set at some convenient value, say 1 volt/div. We will hold the amplitude of the input sine wave constant and vary the frequency. If the scope were an ideal measuring instrument, the peak vertical deflection should (increase, decrease, remain constant) _____ as f is increased.   (a)

Since the scope is not ideal, the peak deflection may vary as f is varied even if the peak input voltage is constant. The peak response of a scope to the input A cos (2πft) is plotted below. If we want to use this scope to make <u>accurate</u> voltage measurements, the maximum frequency which can be used is approximately _____.   (b)

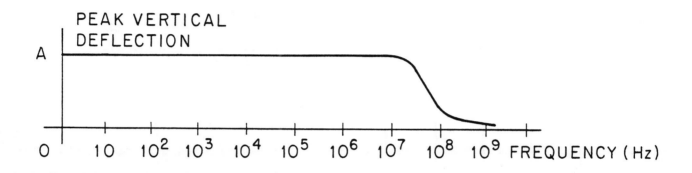

(a) v_1+ to CH1 and to EXTernal trigger input

(b) v_2 to +INPUT (CH1) and −INPUT (CH2); v_1+ to EXTernal trigger input; function generator ground to scope ground in both cases

---

4.29 The phase angle between $v_1 = \sin \omega t$ and $v_2 = \sin(\omega t + 140°)$ is to be measured using the triggered sweep method. Sketch the expected waveforms for step 1 and step 2. Use 20°/division.

Step 1:

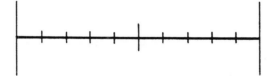

Step 2:

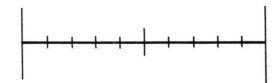

For both steps, the trigger controls should be set as follows:

SOURCE _____   SLOPE _____   LEVEL _____

limiting values: when cos ωt = -1, x = y = -4
when cos ωt = +1, x = y = +4

so draw a diagonal line from (x = y = -4) to (x = y = +4)

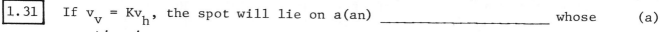

| 1.31 | If $v_v = Kv_h$, the spot will lie on a(an) _____ whose (a)

equation is y = _____. (b)

Sketch the trace if $v_v = 2 \sin ωt$, $v_h = -4 \sin ωt$.

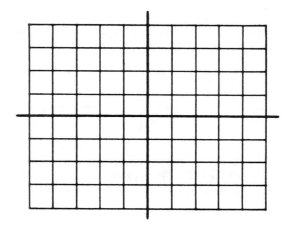

(c)

━━━━━━━━━━━━━━━━━━━━━━━━━━━━━━━━

(a) INTernal   (b) EXTernal (or EXT/10)   (c) 10   (d) no

| 2.32 | When the SOURCE is set to LINE, the triggering signal is the 60Hz AC line.

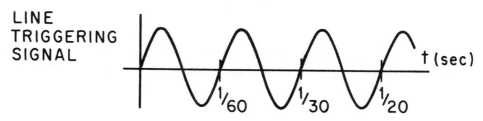

LINE
TRIGGERING
SIGNAL

t (sec)

If the sweep duration, T, is less than 1/60 sec, the sweep will be

triggered _____ times per second. If 1/60 < T < 1/30, the (a)

sweep will be triggered _____ times per second. The sweep (b)

will be triggered 20 times per second, if T lies in the range

_____ < T < _____. (c)

(a) remain constant   (b) 10^7 Hz (10MHz)

3.32 The response of the scope at low frequencies depends on whether the input switch is set to AC or DC. Figs. 3-32a and 3-32b show the response curves of the same scope for two different settings of the input switch. Label the curves with the appropriate switch setting.   (a)

If AC input is selected, the frequency range in which accurate voltage measurements can be made is about _____ to (b)

_____.

If we want to measure the peak value of a 10 Hz sine wave _____ input should be used.   (c)

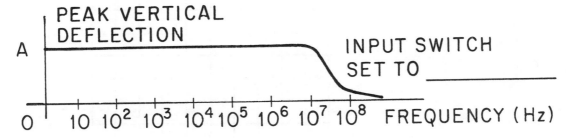

FIG. 3-32a

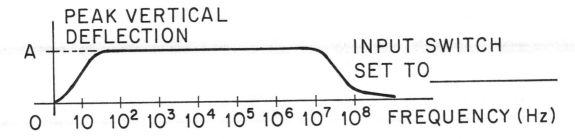

FIG. 3-32b

TURN TO FRAME 4.30

(a) line   (b) y = Kx

(c) a diagonal line between (x = -4, y = +2) and (x = +4, y = -2)

1.32 If $v_h = 4 \cos \omega t$ and $v_v = 4 \sin \omega t$, plot the location of the spot for $\omega t = 0, \pi/4, \pi/2, 3\pi/4, \ldots, 2\pi$. As the spot moves, it traces out a(an) _____. Put an arrow on the plot which indicates the direction of motion of the spot.

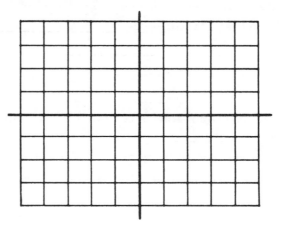

(a) 60   (b) 30   (c) 1/30 < T < 1/20

2.33 If the sweep waveform and vertical input are as shown below, would a stable picture be obtained on the screen? _____.
Explain _____

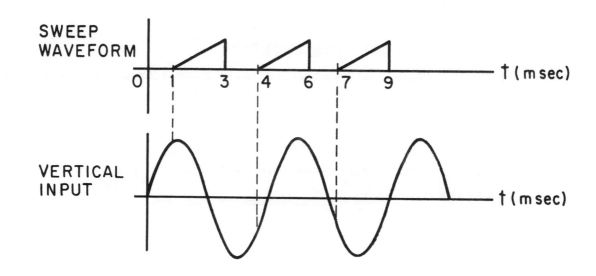

(a) 3-32a is DC input and 3-32b is AC input

(b) 20 Hz to 10^7 Hz   (c) DC

3.33 We have studied several types of errors which can reduce the accuracy of measurements made with the scope. For each type of error, state what can be done to reduce the effects of the error.

Loading error. _____ (a)

_____

Parallax error. _____ (b)

_____

Reading error. _____ (c)

_____

The calibration error which is inherent in the scope may be as large
as _____.   (d)

Errors may also occur if the frequency of the input signal is

_____.   (e)

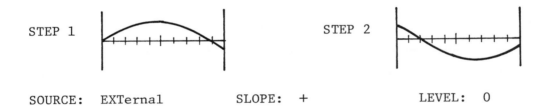

SOURCE: EXTernal   SLOPE: +   LEVEL: 0

THIS IS A GOOD PLACE TO TAKE A BREAK

Phase Shift Measurement by the Ellipse Method

4.30 We will now discuss the ellipse method for phase shift measurement with the scope. As in the triggered sweep method just discussed, the object of the method is to determine the phase angle between sinusoidal voltages, v_1 and v_2. Instead of displaying v_1 and v_2 as functions of time, we will display v_2 vs v_1.

Which control(s) on your scope should be set to which position

to display v_2 vs v_1? _____

circle (of radius 4)
spot moves counterclockwise

1.33 In the preceding frame, the vertical deflection is y = 4 sin ωt and
the horizontal deflection is x = 4 cos ωt. Verify analytically that
the trace is a circle by eliminating the parameter ωt from the equations.

(a)

Hint: $\sin^2 \omega t + \cos^2 \omega t = 1$

Will there be any change in the appearance of the trace if the x and
y inputs are interchanged? _____

(b)

NO. The sweep starts at a different point in each cycle.

2.34 The vertical input and the sweep are said to be synchronized if the
sweep is triggered at the same point in every cycle of the vertical
input.

When the vertical input is a periodic waveform, what relation must
exist between the vertical input and the sweep in order to obtain a
stable trace on the screen? _____

(a)

The proper relation will be obtained if the TRIGGER SOURCE is set to

_____.

(b)

(a)　use the probe　　　(b)　view scope directly in front of screen

(c)　use as large a deflection as possible (turn up the gain)

(d)　3% (5%)　(e)　too high (or too low if the AC input is used)

THIS IS A GOOD PLACE TO TAKE A BREAK

Differential Mode Operation

3.34　We shall describe the function of a differential amplifier and explain
how a dual trace scope may be used in the differential amplifier mode.
The CH1 input on the oscilloscope will be used as the "+INPUT" and the
CH2 input will be used as the "−INPUT".　The voltage applied between
the +INPUT and ground will be called v_+ and the voltage applied between
the −INPUT and ground will be called v_-.

A network is connected to the oscilloscope as shown.　For this connection,

v_+ = _____ and v_- = _____.

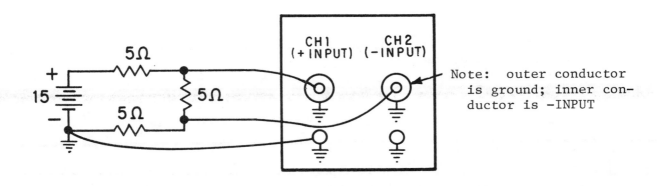

Note:　outer conductor
is ground; inner con-
ductor is −INPUT

T922:　Set SOURCE to X-Y and vertical mode to CH1.

OTHER:　Your scope must be set for X-Y operation, review frame
　　　　1.70 if you need help.

4.31　Indicate the connections to the scope required to display v_2 (on the
vertical) vs. v_1 (on the horizontal) for the network shown below:

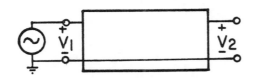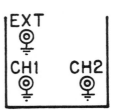

70

(a) Squaring and adding the two equations gives $x^2 + y^2 = 4^2$ which is the equation of a circle of radius 4.

(b) No. (since $y^2 + x^2 = 4^2$ is still a circle of radius 4)

1.34   If $v_h = A \sin \omega t$, in order to obtain a circular trace, v_v must be

_____, but the trace will be a diagonal line   (a)

(not necessarily at 45°) if $v_v =$ _____.   (b)

If $v_h = B \cos \omega t$, in order to obtain a circular trace, v_v must be

_____, but the trace will be a diagonal line (not   (c)

necessarily at 45°) if $v_v =$ _____.   (d)

(a)   vertical input and sweep must be <u>synchronized</u>   (b)   INTernal

---

2.35   A voltage v_x is applied to the external trigger input and v_v is applied to the vertical input. The trigger SOURCE is set to EXT, the SLOPE is set to +, and the LEVEL is set to trigger when v_x is −1 volt. Draw an arrow to indicate the point on v_x at which triggering occurs. Draw an arrow on v_v to indicate the time at which the sweep begins. Then sketch v_v as it will appear on the screen if the sweep rate is 1 ms/div.

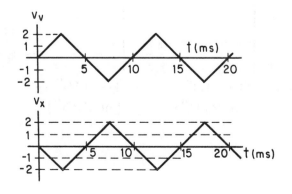
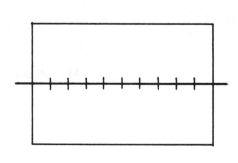

3.35 In the differential amplifier mode, when both inputs are used, the vertical deflection is

$$y = (v_+ - v_-)/S_v$$

If $v_+ = 10$, $v_- = -5$, and $S_v = 5$ volts/div, the vertical deflection

is _____ (a)

If v_+ and v_- are as shown, sketch the trace if $S_v = 2$ volts/div.

(Assume + TRIGGER SLOPE AND 0 TRIGGER LEVEL.)

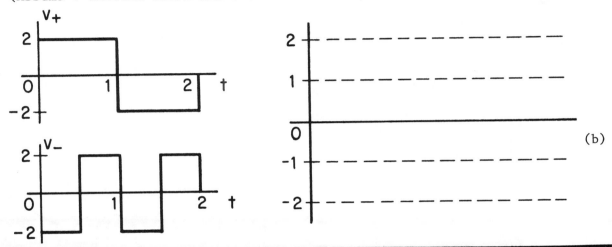

(b)

v_1+ to external horizontal (X) input, v_2+ to vertical (Y) input

v_1- or v_2- to scope ground.

4.32 Assume that $S_v = S_h = 1$ volt/div. If the
horizontal input is $v_1 = A \sin \omega t$ and the
vertical input is $v_2 = A \sin(\omega t + 30°)$
plot the resulting trace on the opposite
graph. (Plot the spot location for $\omega t = 0$,
30, 60, 90, 120, 150, and 180 degrees and
complete the figure by symmetry.)

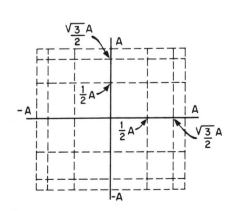

(a) A cos ωt   (b) A_1 sin ωt (A_1 does not have to equal A)

(b) B sin ωt   (d) B_1 cos ωt (B_1 does not have to equal B)

1.35   If the amplitude of the vertical signal used to obtain the circular trace
in frame 1.32 is reduced and the horizontal signal remains the same, sketch
the appearance of the modified trace.

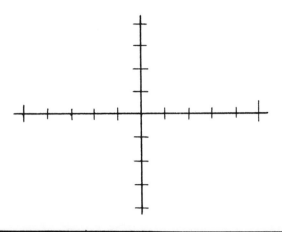

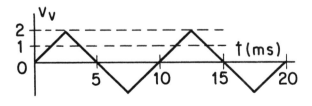

Triggers here   and here

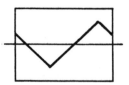

2.36

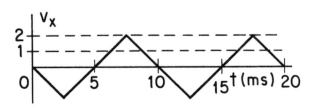

v_v is the vertical input
v_x is the external trigger input
TRIGGER SLOPE set to − (negative)
TRIGGER LEVEL set to trigger
   when the triggering signal
   v_v or v_x is +1 volt
sweep rate = 1 ms/div
Sketch the trace if
(a) TRIGGER SOURCE set to
     INTernal
(b) TRIGGER SOURCE set to
     EXTernal

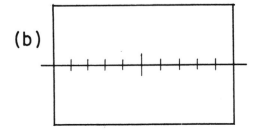

(a)   (b)

(a) $\dfrac{10 - (-5)}{5} = 3$ (b)

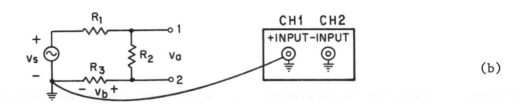

3.36   It is desired to observe v_a with the scope. Terminal 2 cannot be connected to the scope ground because _____ (a)

The voltage from terminal 1 to ground is $v_a + v_b$; the voltage from terminal 2 to ground is v_b. Indicate how to connect the scope so that $v_+ - v_- = v_a$ (b)

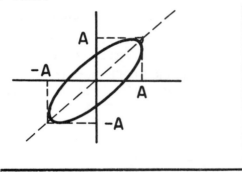

With the above connections, if the sensitivity is S_v, the vertical deflection is $y =$ _____ (c)

Answer to 4.32:

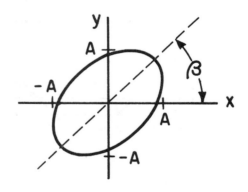

4.33   The angle between the major axis of the ellipse which you just plotted and the x-axis is _____ degrees. (a)

Let $x = A \sin \omega t$ and $y = A \sin(\omega t + \theta)$. The figure at the right gives a plot of y vs x for $\theta = 60°$. Sketch the trace for $\theta = 0$ and $\theta = 90°$ on the same figure. Describe how the x-y plot changes as θ is varied from 0 to $90°$. (b)

_____ (c)

As θ increases, the angle β between the axis of the ellipse and the x-axis (increases/remains the same/decreases)

_____ (d)

1.36 As we have seen, a circular or elliptical trace can be obtained by using a cosine wave and a sine wave as H and V inputs to the scope. The circuit of Fig. 1-36 will be used in lab to produce the required H and V signals.

If we assume that the magnitude of the impedance of the capacitor, $1/\omega C$, is much less than R, the voltage _____ is negligible compared with the voltage v_2, and (a)

$v_2 \approx$ _____. (b)

$i = v_2/R \approx$ _____ (c)

= _____ $\cos \omega t$ (d)

The voltage across C is then

$v_v = \frac{1}{C} \int i \, dt \approx$ _____ $\sin \omega t$ (e)

(substitute for i and carry out the integration)

This approximation is valid if

$1/\omega C \ll R$ or ωRC _____ 1 (f)

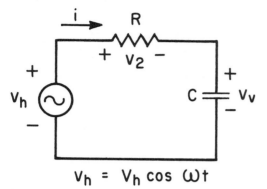

$v_h = V_h \cos \omega t$

FIG. 1-36

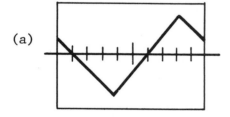

(a)   (b)

(triggers when $v_v = 1$ with neg. slope) (triggers when $v_x = 1$ with neg. slope)

2.37 If the frequency of the vertical input is 70 Hz, could a stable picture be obtained if the TRIGGER SOURCE is LINE? _____ (a)
Explain _____

With the same vertical input, if the frequency of the trigger input is 35 Hz, could a stable picture be obtained if the TRIGGER SOURCE is EXTernal? _____ Explain _____ (b)

_____

(a) R_3 would be shorted out (v_a would change)

(b) terminal 1 to +INPUT, terminal 2 to -INPUT (center terminals)

(c) $y = v_a / S_v$

3.37 Follow these rules when connecting a circuit to the scope:

Rule 1. If some point in the circuit is grounded, connect the circuit ground to the scope ground. Do not rely on the line cord grounds for a connection.

Rule 2. If a voltage to be connected to the scope has one side grounded, then use only the CH1 input. (Select the CH1 vertical mode.)

Rule 3. If a voltage to be connected to the scope has neither side in common with the circuit ground, use both the CH1 (+INPUT) and CH2 (-INPUT) on the scope, and select the differential amplifier mode.

Indicate proper connections to the scope to observe v_1 for each of the following circuits:

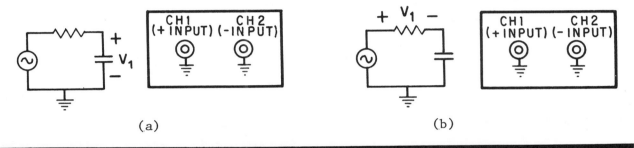

(a)   (b)

TURN TO FRAME 4.34

1.37   As we have just shown, the output of the network of Fig. 1-36 is

$$v_v = (V_h/\omega RC)\sin \omega t$$

A student connects the network of Fig. 1-36 to the scope as shown below, with $v_h = 4\cos \omega t$ ($V_h = 4$) and $\omega RC = 10$. The scope settings
are $S_v = S_h = 1$ volt/div. He observes an ellipse on the screen. In
order to obtain a circular trace of radius 4 divisions, which of the
scope settings should be changed? _____   (a)
To what value? _____   (b)

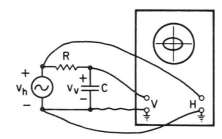

(a) No. The line frequency is 60 Hz so the sweep would not be synchronized with the vertical input.

(b) Yes.   35 Hz is exactly half of 70 Hz.

2.38   The sweep will <u>not</u> trigger at a point of zero slope on the triggering
signal. Suppose that v_1
and v_2 are available as
inputs and we wish to
display the following on
the screen:

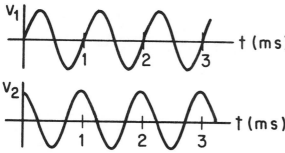

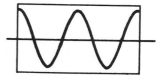

v_1 should be connected to the _____   (a)
v_2 should be connected to the _____   (b)
The controls should be set as follows:
TRIGGER SOURCE _____   SLOPE _____   LEVEL _____   (c,d,e)
SEC/DIV _____   (f)

(a) circuit ground to scope ground; + side of v_1 to +INPUT

(b) circuit ground to scope ground; + side of v_1 to +INPUT;
    − side of v_1 to −INPUT

3.38 Given two signals, v_+ and v_-, the difference component is defined as

$$\Delta v = v_+ - v_-$$

The component which is common to both signals, or common-mode component, is

$$v_c = \frac{1}{2} (v_+ + v_-)$$

For each of the following pairs of signals, find the difference component and the common-mode component:

v_+	v_-	difference component	common-mode component	
5	1	_____	_____	(a)
$2 + 3 \sin \omega t$	$-2 + 3 \sin \omega t$	_____	_____	(b)
$2 \cos \omega t$	$4t$	_____	_____	(c)

(a) 45   (b) 45° diagonal line and circle of radius A

(c) the plot changes from a straight line to an ellipse to a circle (the ellipse grows fatter as θ increases)

(d) remains the same (The major axis is always at a 45° angle with the x-axis when $0 < \theta < 90^{\circ}$ if the maximum values of x and y are the same.)

4.34 The observed trace for $x = A \sin \omega t$ and $y = B \sin(\omega t + \theta)$ is given below for two different values of B. The angle β between the major axis of the ellipse and the x-axis depends only on (check one):

   the value of θ _____   the relative values of A and B _____   (a)

   the angle β is 45° only if _____.   (b)

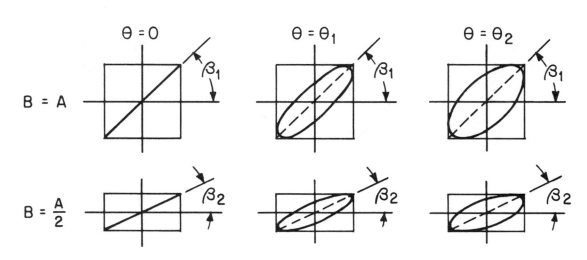

(a) S_v (vertical)   (b) 0.1 volt/div (since $v_v = .4 \sin \omega t$ and we want

a maximum deflection of 4 divisions)

THIS IS A GOOD PLACE TO TAKE A BREAK

Signals with AC and DC Components

1.38

You have now learned the relation between the voltages applied to the input terminals of the scope and the resulting deflection of the spot on the screen. In some cases the input voltage to the scope will have both AC and DC components. In the next sequence of frames, you will learn how to set the scope so that it will respond only to the AC component instead of to the entire input voltage.

By the DC component of a periodic voltage waveform we mean its average value. The average can be taken over one period. Give the DC component of each waveform in Fig. 1-38.

(a) _____   (b) _____   (c) _____

(d) _____

Hint: Parts (a), (b) and (c) can be done by inspection. Use integration for Part (d).

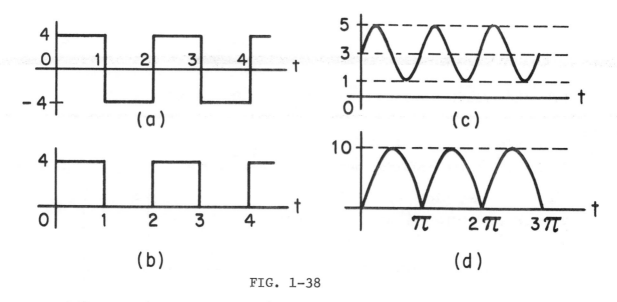

FIG. 1-38

TURN TO FRAME 2.39

(a)   4   3   (b)   4   3 sin ωt

(c)   2 cos ωt – 4t   cos ωt + 2t

3.39 In the differential amplifier mode, if the inputs are v_+ and v_-, the difference component of the input voltage is Δv = _____ and the   (a)

common-mode component is v_c = _____   (b)

An <u>ideal</u> differential amplifier amplifies the difference component and rejects the common-mode component. If the vertical sensitivity is S_v, the vertical deflection produced by an <u>ideal</u> differential amplifier

is   y = _____   (c)

The difference mode gain is defined as $K_d = 1/S_v$. For the ideal differential amplifier, express y in terms of the difference mode gain.

    y = _____   (d)

(a)   the relative values of A and B   (b)   A = B

4.35 x = A sin ωt   y = A sin(ωt + θ)

If θ = 180°, the relation between x and y is y = _____.   (a)

Sketch the x-y plot for θ = 180°.   (b)

What do you think the plot will look like if θ is between 90° and 180°?

Sketch your answer for θ about 135°.   (c)

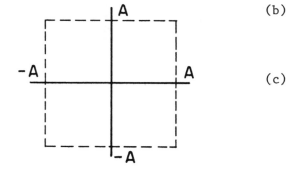

(a) 0   (b) 2   (c) 3   (d) $\frac{1}{\pi} \int_o^{\pi} 10 \sin t \, dt = \frac{20}{\pi} = 6.37$

1.39   To obtain the AC component of a periodic voltage waveform, subtract the DC component from the total voltage.

Sketch the AC component of each of the waveforms shown in Fig. 1-38. Specify the peak values of the AC component.

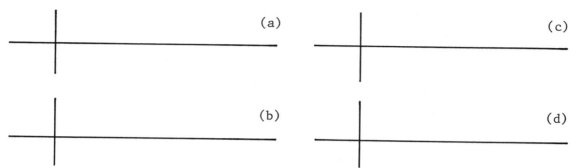

Note that for the AC component, the area below the axis is equal to the area above the axis, so the average value of the AC component is always _____.
                                                                          (e)

Two different solutions are possible:

(a) vertical input   (b) ext. trigger input   (c) EXT.   (d) –
(e) 0   (f) .2 ms/div.
OR
(a) ext. trigger input (b) vertical input   (c) EXT.   (d) +
(e) 0   (f) .2 ms/div.

2.39   When observing a periodic signal with the scope, the proper TRIGGER SOURCE must be selected in order to obtain a stable picture. Complete the following table giving all allowable settings of the TRIGGER SOURCE switch which would give a stable trace:

Vertical Input	External Trigger Input	Allowable Settings of Trigger Source	
120Hz	NONE		(a)
56Hz	28 Hz		(b)
140Hz	60 Hz		(c)
1190Hz	DC		(d)

(a) $v_+ - v_-$ (b) $\frac{1}{2}(v_+ + v_-)$ (c) $\frac{1}{S_v}(v_+ - v_-)$

(d) $K_d(v_+ - v_-)$

3.40   Several difficulties may be encountered in the differential mode of
operation because the differential amplifier is not ideal.   Instead
of producing a vertical deflection of

$$y = K_d(v_+ - v_-)$$

the actual deflection is

$$y = K_d(v_+ - v_-) + K_c \frac{v_+ + v_-}{2} = K_d \Delta v + K_c v_c$$

K_d is called the _____ gain and   (a)

K_c is called the common-mode gain.

If $y = 10\Delta v + .1 v_c$, the common-mode gain is _____ and the   (b)

difference mode gain is _____ .   (c)

For an ideal differential amplifier, $K_c = $ _____   (d)

(a) $y = -x$

(b,c)

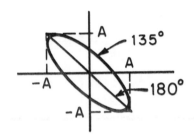

4.36   Sketch the location of the major axis of the ellipse

(a)   if $0 \le \theta < 90^\circ$

(b)   if $90^\circ < \theta \le 180^\circ$

If the ellipse is very thin, θ is close to _____ or _____   (c)

If the ellipse is very fat, θ is close to _____ .   (d)

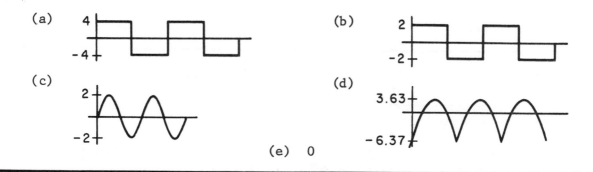

(a) 4 -4   (b) 2 -2

(c) 2 -2   (d) 3.63 -6.37

(e) 0

1.40 State the relation between a voltage v(t), its DC component V_{DC}, and its AC component $v_{AC}(t)$.

(a) INT, LINE   (b) INT, EXT   (c) INT   (d) INT

THIS IS A GOOD PLACE TO TAKE A BREAK

Trigger Coupling

2.40 A typical coupling circuit for the input to the triggering circuit is shown below. The triggering signal is direct coupled to the input of the trigger pulse generator when the COUPLING switch is set to _____. (a)
The signal is coupled through a blocking capacitor when the switch is set to _____. The DC level of the triggering signal does not (b)
affect the sweep circuit when the coupling is set to _____. (c)

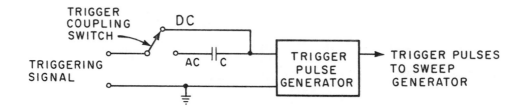

(a) difference mode   (b) .1   (c) 10   (d) 0

3.41 The vertical deflection produced by a non-ideal differential amplifier

is y = _____ (a)

(Express your answer in terms of K_d, K_c, v_+ and v_-)

The <u>common-mode rejection ratio</u>* (CMRR) is defined as the ratio of
the difference mode gain to the common mode gain:

$$CMRR = K_d/K_c$$

If y = 15 Δv + .2 v_c, the common-mode rejection ratio is _____ (b)

If the differential amplifier is ideal, the common-mode gain is ____, (c)
so the common-mode rejection ratio is _____. (d)
A good differential amplifier will have a (high/low) _____ CMRR. (e)

_____

*also called differential rejection ratio

(a)

(b)

(c) 0° or 180°

(d) 90°

4.37 If we change the sign of θ from + to -, the x-y plot will be unchanged.
(Proof of this statement is given in the next frame.)

Is it possible to distinguish between θ = 60° and θ = -60° using the
ellipse method? _____ (a)

Between θ = -30° and θ = 330°? _____ (b)

A student determines that θ = 120° (or 240°) by the ellipse method.
What other value might θ have? _____ (c)

What method of phase measurement could be used to determine which of
the two values is correct?

_____ (d)

$$v(t) = V_{DC} + v_{AC}(t)$$

1.41 In Fig. 1-41a, the steady-state DC current flowing through C and R is

i = _____ and the steady-state DC voltage across R is (a)

v_R = _____. The steady-state DC voltage across the capaci- (b)

tor is v_C = _____. (c)

In Fig. 1-41b, assume that the impedance of the capacitor is very

small compared with R at all frequencies present in $v_{AC}(t)$. The

voltage across the resistor is v_R = _____. (d)

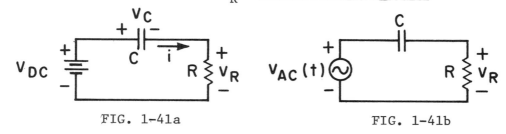

FIG. 1-41a   FIG. 1-41b

In this example, the capacitor acts like a(an) _____ (e)

circuit for DC and a(an) _____ circuit for AC. (f)

(a) DC   (b) AC   (c) AC

2.41 If we want the triggering circuit to respond to the entire input wave-

form including any DC component, the trigger coupling should be set

to _____. (a)

If we want the triggering circuit to respond to only the time-varying

component of the input, we should set the trigger coupling to

_____. (b)

(a) $K_d(v_+ - v_-) + K_c \dfrac{v_+ + v_-}{2}$ (b) $15/.2 = 75$

(c) 0   (d) infinity   (e) high

3.42   In order to measure the CMRR, we must determine K_d and K_c. Given

that $y = K_d(v_+ - v_-) + K_c \dfrac{v_+ + v_-}{2}$ what is the deflection if

$v_+ = v_- = v_1$? $y =$ _____   (a)

Indicate how we should connect an oscillator to determine K_c.

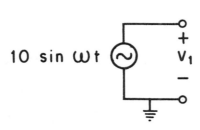
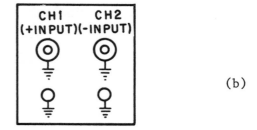

                                                                    (b)

With this connection, if the peak vertical deflection is 1 division

     $K_c =$ _____ div/volt   (c)

(a) NO (since changing the sign of θ does not change the x-y plot)

(b) NO (330° is the same as -30°)

(c) -120° (or -240°)

(d) dual-trace (or triggered sweep) method

4.38   Proof follows that changing the sign of the phase angle does not change the ellipse which is displayed. (You may skip the rest of this frame if you wish.)

Assume that the horizontal deflection is $x = A \sin \omega t$ and the vertical deflection is $y = B \sin(\omega t + \theta)$. At time t_1, the spot is located at $x_1 = A \sin \omega t_1$, $y = B \sin(\omega t_1 + \theta)$. If we change the y deflection to

     $y = B \sin(\omega t - \theta)$, at time $t_2 = (\pi/\omega) - t_1$ we have

     $x = A \sin \omega t_2 = A \sin(\pi - \omega t_1) = A \sin \omega t_1 = x_1$

   and $y = B \sin(\omega t_2 - \theta) = B \sin(\pi - \omega t_1 - \theta) = B \sin(\omega t_1 + \theta) = y_1$

Hence for every point (x_1, y_1) on the original ellipse, there is a corresponding point on the ellipse which is obtained when the sign of θ is changed. In one complete cycle, the same set of points is traced out in both cases, so the resulting ellipses are identical.

(a) i = 0 (since DC current cannot flow through a capacitor)

(b) v_R = iR = 0   (c) v_C = V_{DC} (since there is no voltage across R)

(d) v_R = $v_{AC}(t)$ (since the voltage drop across C is negligible)

(e) open (for DC)   (f) short (for AC)

1.42   In the circuit below, what is the DC component of $v_1(t)$? _____ (a)

If the impedance of the capacitor is very small compared with R at all

frequencies present in $v_{AC}(t)$, $v_1(t)$ = _____ (b)

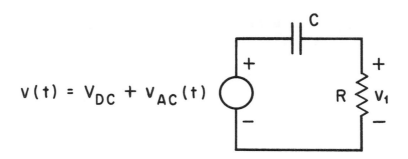

(a) DC   (b) AC

2.42   When observing very low frequency signals, the AC-DC-GND switch on

the vertical amplifier should be set to _____. (a)

In a similar manner, for very low frequency triggering signals (below

16Hz) the TRIGGER COUPLING must be set to _____. (b)

Given that the triggering signal is 5 + 2 sin 2πt, is it possible to

set the trigger coupling so that the trigger circuit only responds to

the AC component (2 sin 2πt) and <u>not</u> to the DC component (5)? _____ (c)

Explain. _____.

(a) $K_c v_1$ (c) 1 div / 10 volts = .1 div/volt

(b) connect + side of v_1 to both +INPUT and −INPUT (center terminals)
     connect the oscillator ground to the scope ground (with this connec-
     tion v+ = v− = v_1)

3.43   In the differential amplifier mode, v_1 = 5 sin ωt is connected to both the
       +INPUT and the −INPUT. A peak deflection of 2 divisions is observed, so

       K_c = _____.   (a)

       The scope is set to S_v = .01 volts/div, so K_d = _____.   (b)
       Therefore, the common-mode rejection ratio is _____.   (c)

4.39   Sketch the trace approximately for each of the following values of θ:

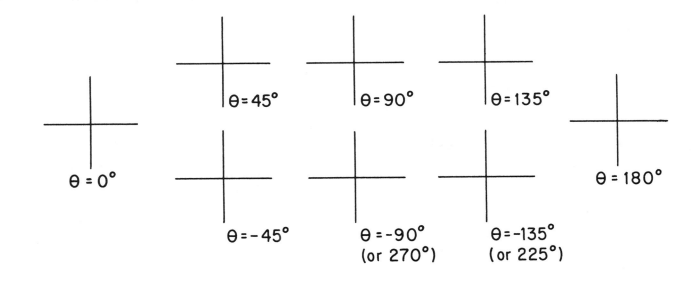

(a) 0 (since no DC current can flow through the capacitor)

(b) $v_{AC}(t)$ (since the AC voltage across the capacitor is negligible)

1.43 In the preceding circuit, C is referred to as a blocking capacitor because it blocks the _____ and passes only (a)
the _____. (b)

(a) DC   (b) DC

(c) No, because DC coupling must be used for signals below 16 Hz.

2.43 If TRIGGER LEVEL = 0, TRIGGER SLOPE = +, circle and label the points on the following waveform at which triggering will occur (a) for AC coupling and (b) for DC coupling.

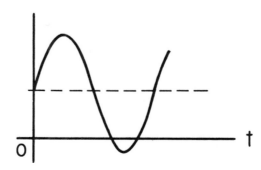

Note: In this and the following figures, the dashed line is the average value (DC value) of the waveform. If you need to review AC and DC components, turn to page 236.

(a) 2/5 = .4 div/volt   (b) 1/.01 = 100 div/volt

(c) 100/.4 = 250

3.44 The following procedure can be used to determine the common-mode rejection ratio:

1. Adjust the oscillator output for a peak voltage V_p.

2. Then connect the oscillator to the scope and observe the peak deflection y_1 for a sensitivity setting S_v. Show the required connections:

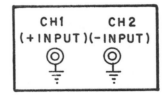   (a)

3. Calculate the CMRR in terms of V_p, y_1, and S_v using the following equations:

_____   _____   _____   (b)

4.40 Fill in the number of the ellipse which corresponds to each value of θ:

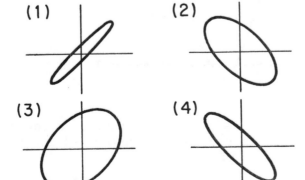

θ	ellipse number	
+ 15		(a)
+ 70		(b)
+ 155	*	(c)
+ 130	*	(d)

*Be careful. Think before you write your answer.

(a) DC component   (b) AC component

1.44 The scope amplifiers are direct-coupled and pass signals of all frequencies from DC to a certain maximum frequency limit. In some applications, the signal to be observed with the scope has both AC and DC components. If we are only interested in observing the AC component, it is necessary to block the _____ with a   (a)

_____.   (b)

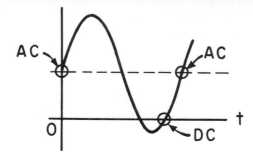

2.44 Do not confuse the effect of the AC-DC vertical input switch with the effect of the AC-DC trigger coupling switch.

Trigger source is set to INT.

The following waveform appears on the screen with the vertical input set to DC and with AC trigger coupling:

For this observed waveform the trigger slope is _____   (a)
and the trigger level is

_____.   (b)

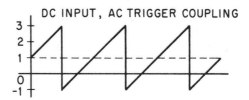

The vertical input is now changed to AC. The sweep will trigger at the same point in time because _____   (c)

_____

Sketch the new waveform observed on the screen.   (d)

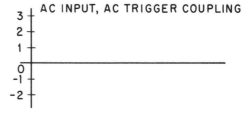

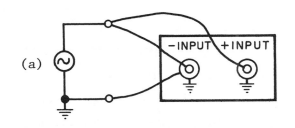

(a)

(b) $K_d = 1/S_v$, $K_c = y_1/V_p$

$$CMRR = K_d/K_c = V_p/y_1S_v$$

3.45 If $S_v = .5$ and CMRR = 1000

$K_d = $ _____ and $K_c = $ _____   (a,b)

If $v_+ = 8$ and $v_- = 4$, the difference mode component is

$\Delta v = $ _____ and the common-mode component is $v_c = $ _____   (c,d)

The actual deflection would be _____ compared with   (e)

_____ for an ideal differential amplifier. The per cent   (f)

error in the deflection would be _____.   (g)

(a) (1)   (b) (3)   (c) (4)   (d) (2)

4.41 What is the possible range for the magnitude of θ if the major axis

of the ellipse is tilted this way: ? _____< $|\theta|$ < _____   (a)

If the major axis is tilted the other way? _____< $|\theta|$ < _____   (b)

Note: $|\theta|$ is always a positive number. If $|\theta| = 15^\circ$, then

$\theta = $ _____ or _____.   (c)

(a) DC component   (b) blocking capacitor

1.45 A typical input arrangement to the vertical amplifier is shown in
Fig. 1-45. R represents the impedance seen at the input terminals of
the vertical amplifier and C is a blocking capacitor. When the AC
input is selected, the scope input is coupled to the amplifier input
through a _____.   (a)

If S_v = 1 volt/div, the DC input is selected, and the scope input

voltage is v_v = 3 volts DC, the vertical deflection will be

_____ divisions.   (b)

If the input is now switched to AC input, the blocking capacitor will

charge up to _____ volts, v_v' will become _____ volts,   (c,d)

and the vertical deflection will be _____ divisions.   (e)

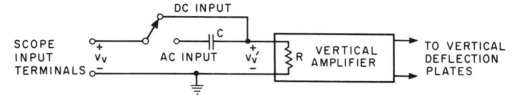

FIG. 1-45

(a) +   (b) 0

(c) the AC component is the   (d)
     same in both cases

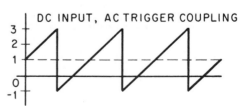

2.45 The following waveform is observed
with the vertical input set to DC
and with AC trigger coupling.

Now the trigger coupling is switched
to DC (leaving everything else
unchanged). The waveform does <u>not</u>
shift up or down because

_____ (a)

_____

The waveform <u>does</u> shift sideways
because _____

_____ (b)

Sketch the new waveform as observed
on the screen.   (c)

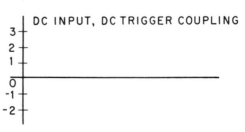

(a) 1/.5 = 2   (b) 2/1000 = .002   (c) 8 - 4 = 4

(d) $\frac{1}{2}$(8 + 4) = 6   (e) $K_d \Delta v + K_c v_c = 2 \times 4 + .002 \times 6 = 8.012$

(f) 8   (g) (.012/8) x 100% = 0.15%

3.46 If v_+ = 10.05, v_- = 10.00, S_v = .1 volt/div, and the common-mode rejection ratio is 1000, the vertical deflection is

_____. (a)

The per cent error in the deflection is _____. (b)

When v_+ and v_- are very close together, a large error in the deflection is possible. The larger the common-mode rejection ratio, the _____ will be the error. (c)

(a) $0° < |\theta| < 90°$ (b) $90° < |\theta| < 180°$ (c) $+15°$ or $-15°$

4.42 We will now derive a method for finding the numerical value of θ from the ellipse.

Let x = A sin ωt, y = B sin(ωt + θ)

d_1, the peak-to-peak vertical deflection, is _____ (a)

Where the ellipse intersects the y-axis, x = A sin ωt = 0, so ωt = 0 or 180°, and y = _____ or _____. (b)

d_2, the distance between the y-axis intercepts is _____ (c)

   d_2/d_1 = _____ = _____ (d)

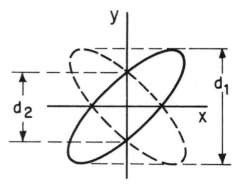

(a) blocking capacitor (b) 3 (c) 3

(d) 0 (since no DC current is flowing, there is no voltage across R)

(e) 0

1.46 In the figure below, suppose that $v_v = 10 + 5 \sin \omega t$. If the DC
(direct-coupled) input is selected, $v_v' =$ _____ (a)
Assume that ω is high enough so that the AC impedance of the capacitor
is negligible compared with R. If the AC input is selected, C will
charge up to 10 volts, and $v_v' =$ _____. (b)
If $S_v = 5$ volts/div and the AC input is used, the vertical deflection
will be between the limits _____ and _____. (c)
If $S_v = 5$ volts/div and the DC input is used, the vertical deflection
will be between the limits _____ and _____. (d)

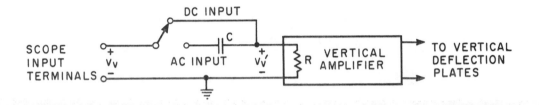

(a) the DC component of the input
    is unchanged

(b) the sweep triggers at a different
    time

(c)

2.46 The spot will stay at the left edge of the screen until the sweep is triggered. When the spot is not moving across the screen, the intensity is
automatically reduced so the spot cannot be seen. We say that the spot is
"blanked out" when it is not moving across the screen from left to right.

For the sweep waveform shown below, plot the corresponding spot intensity
as a function of time. (Your plot should show at what times the spot is
normal intensity and at what times it is blanked out.)

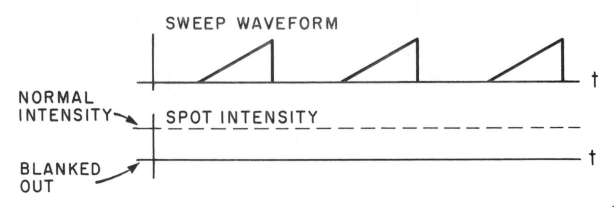

(a)   .6 div   (b)   20%   (c)   smaller

3.47   Even when the difference of the two inputs is small, the maximum voltage which can be applied to either the +INPUT or the -INPUT must be
limited to avoid overloading the scope amplifier and distorting the
signal. Generally, the maximum peak input voltage per channel, with no
distortion, is about 20 times the VOLTS/DIV setting.
If S_v = 10 mv/div., what is the maximum allowable input voltage?

_____   (a)

When only one input is used, if the vertical deflection of the spot is
such that it remains on the screen we know that the maximum input
voltage has not been exceeded. Is this statement true when both + and
- inputs are used? _____. Explain. _____ (b)

_____

_____

(a)   2B (since the peak value of y is B)   (b)   B sin θ or -B sin θ

(c)   2B sin θ   (d) $\dfrac{2B \sin \theta}{2B}$ = sin θ

4.43   Give a formula for computing $|\theta|$ in terms of distances which can be
measured on the ellipse. $|\theta|$ = _____   (a)
Show these distances on the ellipses below.

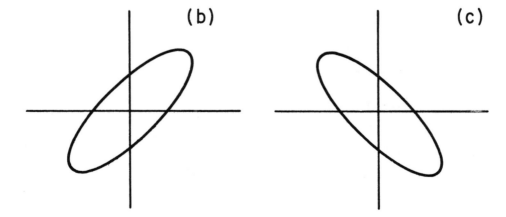

(b)   **(c)**

96

(a) $10 + 5 \sin \omega t$ (b) $5 \sin \omega t$ (c) +1 and -1 div

(d) +3 and +1 div (since maximum v_v is 10 + 5(1) and minimum v_v is 10 + 5(-1)

1.47   If the input signal contains both AC and DC components, we should use the AC input if we want to observe the _____ only,   (a)

but we should use the DC input if we want to observe both the

_____.   (b)

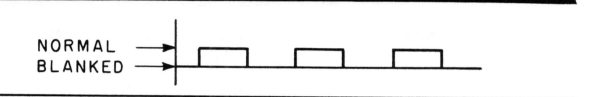

2.47   At very high sweep rates, the time required to reset the sweep generator is no longer neglibibile as it is at slower sweep rates. The sweep waveform and the corresponding spot intensity are shown below for a very high sweep rate:

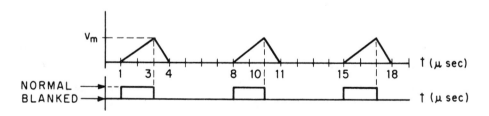

If v_m is the voltage required to deflect the spot 10 divisions, what is the sweep rate? _____ μ sec/div*   (a)

In the above example, how much time does it take to reset the sweep generator to 0? _____   (b)

Can the spot be seen when it is returning from the right side of the screen to the left? _____   (c)

---

*The symbol μ means "micro" or 10^{-6}.

(a)   ≈ 200 mV (10 mV x 20)

(b)   No. The difference of the inputs and the deflection may be small even if the maximum input voltage is exceeded.

3.48   If the maximum input voltage is exceeded at either input, the signal may be badly distorted. If $v_+ = 8 \sin \omega t$ and $v_- = 8.5 \sin \omega t$, the smallest VOLTS/DIV setting which can be used without danger of distorting the signal is _____ . The corresponding peak-to-peak   (a)
deflection is _____ . (Assume that the common-mode   (b)
rejection ratio is very high.)

Note: peak-to-peak is two times peak.

(a) $|\theta| = \sin^{-1}(d_2/d_1)$

(b,c) d_1 is the distance between the top and bottom of the ellipse.
d_2 is the distance between the y-axis intercepts.

4.44   The above formula gives a value of $|\theta|$ between 0 and 90° and a second value between 90 and 180°. How do we tell which value is correct?

_____   (a)

If we change the vertical gain of the scope, will this change the value of $|\theta|$ which is computed by the formula? _____

Explain. _____   (b)

(a) AC component   (b) AC and DC components

1.48 | At very low frequencies, the impedance of the blocking capacitor
$(1/\omega C)$ is <u>not</u> negligible compared with the input impedance of the
amplifier (R). Therefore, at very low frequencies, the voltage drop

across the _____ is not negligible, and the   (a)

AC input (should/should not) _____ be used.   (b)

(a) $\frac{2\mu sec}{10\ div}$ = .2 μsec/div

(b) 1 μsec (between 3 and 4 μsec, 10 and 11 μsec, etc.)

(c) NO. (The spot is blanked out during the time the sweep generator is
         resetting.)

2.48 | The TRIGGER MODE control has two (or more) positions. In the NORMAL
position, the sweep is triggered whenever a triggering signal is received
and the scope has completed the previous sweep. In the AUTO (automatic
triggering) position, the sweep will be triggered automatically at
regular intervals even when no triggering signal is present.
If the TRIGGER SOURCE is set to INTernal and no vertical input signal is
present, what should be observed on the screen if the TRIGGER MODE is

set to NORMAL? _____   (a)

If the MODE is now switched to AUTO, what should be observed at low

sweep rates? _____   (b)

At high sweep rates? _____   (c)

(a) .5 (.2 volts/div couldn't be used since the input is greater than 5 volts)

(b) 2(8.5 - 8)/.5 = 2 div.

3.49 The scope is used in the differential amplifier mode with

+INPUT = .2 sin 100t + .5 cos 1000t

−INPUT = .2 sin 100t

If the volts/div setting is .1, what will the vertical deflection be if the amplifier is ideal?   y = _____ (a)

What will it be if the common-mode rejection ratio is 100?

y = _____ (b)

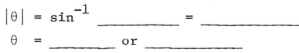

(a) from the slope of the major axis of the ellipse

(b) NO, because d_2 and d_1 change in the same ratio.

4.45 For each of the ellipses below, compute the two possible values of θ.

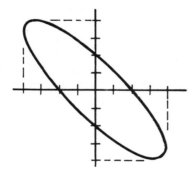 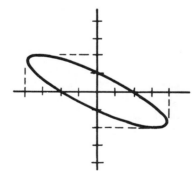

$|\theta| = \sin^{-1}$ _____ = _____

θ = _____ or _____   θ = _____ or _____

            (a)   (b)

100

(a) blocking capacitor   (b) should not

1.49 | If our input signal contains only an AC component and the frequency
is high enough so that the impedance of the blocking capacitor ($1/\omega C$)
is small compared with the input impedance (R), we (should use the
DC input/should use the AC input/can use either the AC or DC input)

_____. At very low frequencies,   (a)

we must use the _____ to observe an AC signal, because   (b)

_____   (c)

(a) Nothing. (The sweep is not triggered. The spot remains at the left
     edge of the screen, but it is blanked out because it is not moving.)

(b) The spot will sweep across the screen.

(c) The moving spot will become a horizonal line.

2.49 | With no external signals connected either to the vertical input or to
the external trigger input, we wish to observe a horizontal line on the
scope. This can be accomplished by setting the TRIGGER MODE to

_____ or by setting the TRIGGER SOURCE to _____. (a,b)

A student connects a sine wave to the vertical input of the scope and
finds that he cannot obtain a stable trace for any settings of the TRIGGER
LEVEL and TRIGGER MODE controls. Which of the triggering control switches

is probably set wrong? _____ To what position should   (c)

it be set? _____   (d)

(a) 5 cos 1000t   (b) 5 cos 1000t + .01 (2.5 cos 1000t + 2 sin 100t)

= 5.025 cos 1000t + .02 sin 100t

3.50 S$_v$ is set to 2 volts/div.   v$_+$ = 60 sin ωt and v$_-$ = 54 sin ωt

The <u>peak-to-peak</u> deflection should be _____ div.   (a)

If the waveform on the screen is distorted, what is the probable

cause? _____ (b)

What could be done to eliminate this distortion? _____

_____   (c)

Other than distortion, what problem might be encountered in using the

differential amplifier if v$_+$ and v$_-$ have a large common mode component

and a small difference component? _____

_____ (d)

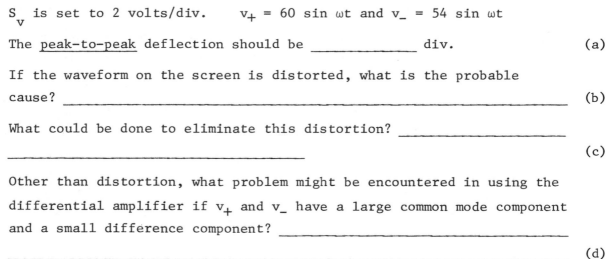

(a) $|\theta| = \sin^{-1} \frac{4}{8} = 150°$ (note that 30° is not correct because of the
way the ellipse is tilted)

θ = 150° or −150°   (since there is no way to tell the sign of
the angle)

(b)   θ = 150° or −150°   (the ratio of d$_2$ to d$_1$ is the same as in (a)).

4.46 When using the ellipse method, the horizontal position must be adjusted
very carefully so that the trace is exactly centered in the horizontal
direction.

Why?_____ (a)

Is it necessary to adjust the vertical position carefully so that the

trace is exactly centered in the vertical direction? _____ (b)

Explain _____

_____

(a) can use either AC or DC input

(b) DC

(c) the impedance of the blocking capacitor is not negligible

---

1.50 In which of the following cases will the trace be an ellipse

(or circle)? _____ (a)

In which cases will the trace by a diagonal line? _____ (b)

case (1): $v_v = 3 \sin \omega t$ $v_h = 4 \cos \omega t$

case (2): $v_v = 2 \cos \omega t$ $v_h = 5 \cos \omega t$

case (3); $v_v = 2 \cos \omega t$ $v_h = 5 \sin \omega t$

case (4): $v_v = 2 \sin \omega t$ $v_h = 4 \sin \omega t$

(a) AUTO   (b) LINE   (c) SOURCE   (d) INTernal

2.50 For the rest of this PART, assume that the TRIGGER MODE switch is in the
NORMAL position.
The trigger controls are set as follows:

LEVEL = 0, SLOPE = +, SOURCE = INTernal,

SEC/DIV = .05 sec/div.

The vertical input is a 1Hz sine wave.

The AC-DC-GND switch should be set to _____. (a)

The TRIGGER COUPLING should be set to _____. (b)

How many cycles will appear on the screen? _____ (c)

If you want 2 cycles of the same sine wave to appear, to what value

should SEC/DIV be changed? _____ (d)

(a)   6 div.   (b)   input voltages v_+ and v_- are too large

(c)   change to 5 volts/div range

(d)   loss of accuracy because common-mode component is amplified and not completely rejected.

3.51   The vertical amplifier is set to .1 volt/div. Probes are connected to both the +INPUT and -INPUT on the differential amplifier, and 10 volts is applied to both probe inputs. If the attenuation of both probes is the same and the differential amplifier is ideal, the vertical deflection will be _____   (a)

Two probes will not have exactly the same attenuation. Suppose that the probe on the +INPUT attenuates the signal by a factor of 9.8 and the probe on the -INPUT by a factor of 10.2. The actual input voltage on the +INPUT will be _____ and on the -INPUT will be _____   (b)

Still assuming the differential amplifier itself is ideal, the vertical deflection will be _____ divisions.   (c)

When the differential amplifier is used in the differential mode with mismatched probes, the effective CMRR (that is, the CMRR measured with respect to the probe input terminals) will be (larger/the same/ smaller) _____ than the CMRR of the amplifier itself. (d)

(a)   The value of distance between the y intercepts would change if the trace was shifted horizontally.

(b)   No. Neither d_1 or d_2 changes when the trace is shifted up or down.

4.47   Compute accurately the value(s) of θ for each of the following ellipses:

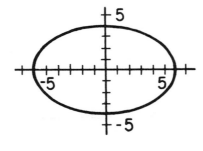

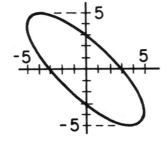

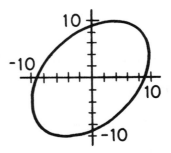

$\theta =$ _____ or _____ $\theta =$ _____ or _____ $\theta =$ _____ or _____

(a)   (b)   (c)

(a)   1 and 3   (b)   2 and 4

1.51 Sinusoidal voltages of the same frequency are applied to both the
vertical and horizontal scope inputs.

If the AC components of v_v and v_h are in phase, the trace will be
a(an) _____.   (a)

If the AC components of v_v and v_h are 90° out of phase (i.e., one is
a sine and one is a cosine), the trace will be a(an) _____

_____.   (b)

(a,b) DC (since the frequency of the signal is very low)

(c)   1/2   (d)   .2 sec/div

2.51 v_1 is connected to the external trigger input and v_2 to the vertical input.
The trigger controls are set as follows: SOURCE = EXTernal, SLOPE = +,
LEVEL = 0, SEC/DIV = .05 ms/div, COUPLING = AC.

The vertical sensitivity is 5 volts/div. Sketch the pattern which appears
on the screen. Work this problem carefully; don't guess. Label the trace
which appears the first time the sweep is triggered and the one which
appears the second time.

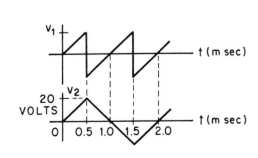

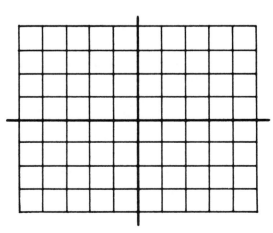

(a) 0 (since the amplifier is ideal and both inputs are the same)

(b) $v_+ = 10/9.8 = 1.02$ $v_- = 10/10.2 = .98$

(c) $y = (v_+ - v_-)/S_v = (1.02 - .98)/.1 = .4$ div

(d) smaller (because the common-mode signal is slightly different at the +INPUT and -INPUT, it is not completely rejected)

3.52 Since the probes in the laboratory are not matched, is it desirable to use the probes when using both the +INPUT and -INPUT on the differential amplifier? _____ Explain. _____ (a)

_____

The most sensitive range on the scope (without the probe) is 2mv/div. If the signal being observed had an amplitude of 2 mv, would it be desirable to use the probe? _____ Explain. _____ (b)

_____

(a) $\pm 90°$ (b) $\pm 143°$ (c) $\pm 64°$

THIS IS A GOOD PLACE TO TAKE A BREAK

The Webb Mask

4.48 Since the thickness (D) of the ellipse varies with the angle θ, it is possible to determine θ from this thickness. Let us work out an example of this. Assume that the horizontal deflection is $x = 4 \sin \omega t$ and the vertical deflection is $y = 4 \sin (\omega t + \theta)$. The resulting trace is shown below. If the thickness is $D = 4\sqrt{2}$, compute the values of x and y which correspond to the point P on the ellipse. (Note that x is negative.)

x = _____

y = _____ (a)

From the above value of x, compute the corresponding value of ωt = _____ (b)

From the above value of y, compute $\omega t + \theta$ and hence determine θ.

$\omega t + \theta$ = _____

θ = _____ degrees   (c)

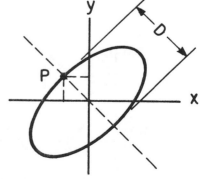

Check your value of θ to see if it is reasonable considering the orientation and size of the ellipse. If it isn't, check your calculations.

106

(a) line

(b) ellipse (or circle if the amplitudes are equal)

1.52 If x = 2 + 3 cos ωt and y = 1 + 3 sin ωt

    maximum x deflection = _____   (a)

    minimum x deflection = _____ (a negative number)   (b)

    maximum y deflection = _____   (c)

    minimum y deflection = _____ (also negative)   (d)

Now sketch a rectangle on the screen showing the limiting values of the x and y deflections. The trace must lie within this rectangle.

Sketch the trace.

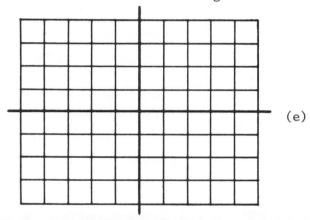

(e)

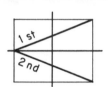

2.52 v_v is applied to the vertical input with the scope set as follows:

vert. input to DC, trigger source to INT, trigger slope to – (negative), trigger coupling to AC, trigger level set to trigger when input voltage equals 1 volt, time/div to 0.5 ms, and volts/div to 1.

(a) Sketch the sweep waveform.   (b) Sketch the waveform which appears on the screen.

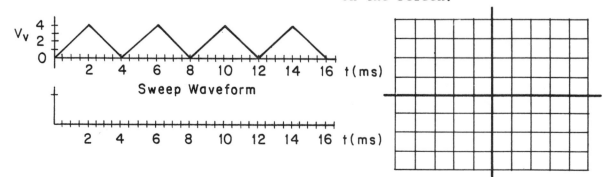

(a) No, because mismatched probes would decrease the CMRR.

(b) No, because the deflection would be reduced to .1 div.

3.53 Check to indicate proper usage of the probe in each of the following situations:

	probe should be used	probe could be used if desired	probe should not be used
(a) low frequency signals			
(b) high frequency signals			
(c) low impedance circuits			
(d) high impedance circuits			
(e) very low amplitude signals			
(f) high amplitude signals			
(g) both +INPUT and −INPUT are being used (and common-mode signal is large)			
(h) only CH1 is being used			

(a) $x = 2\sqrt{2} \cos 135^\circ = -2$ $y = 2\sqrt{2} \sin 135^\circ = 2$

(b) $4 \sin \omega t = -2$, $\omega t = -30^\circ$

(c) $4 \sin(\omega t + \theta) = 2$, $\omega t + \theta = 30^\circ$ $\theta = 60^\circ$

4.49 The Webb mask, which consists of a transparent overlay of the form shown in Fig. 4-49 (next page) provides a convenient way of measuring the thickness of the ellipse. This mask is placed over the scope screen so that the ellipse may be viewed through it. If the <u>major</u> axis of the ellipse lies along line M-M' on the mask, the possible range for $|\theta|$ will be _____ to _____.   (a)

If it lies along N-N' the range for $|\theta|$ will be _____ to _____.   (b)

(In answering this question, ignore the numbers printed on the mask.)

(a) 5   (b) −1
(c) 4   (d) −2   (e)

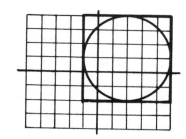

1.53 Let x = −1 + 3 cos ωt   y = 1 + 2 cos ωt

The trace will be a(an) _____ between the points   (a)

x = _____, y = _____   and   x = _____, y = _____   (b)

(a)

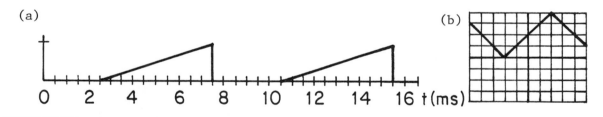

(b)

Dual Trace

2.53 Many oscilloscopes have a dual trace capability that allows you to display
both vertical amplifier Channel 1 and Channel 2 signals on the screen.
As shown below, an electronic switch alternately connects the CH1 and CH2
inputs to the vertical amplifier.
When the switch is in the "upper" position, the scope displays the

_____ input and when the switch is in the "lower" position,   (a)

the scope displays _____.   (b)

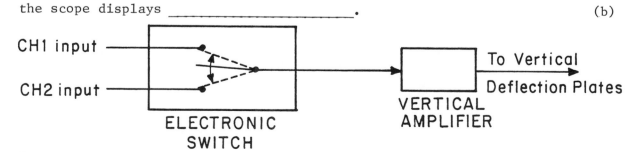

Probe should be used in (b) and (d) to prevent loading the circuit.
Probe should <u>not</u> be used in (e) and (g).

Other cases are optional.

| 3.54 | You have now completed Preparation Part III. You should understand some of the sources of errors which occur when the scope is used to make accurate measurements, and you should know how to avoid these errors. You should know how to use the probe and the differential amplifier mode. Now go to the lab and do Lab Part III.

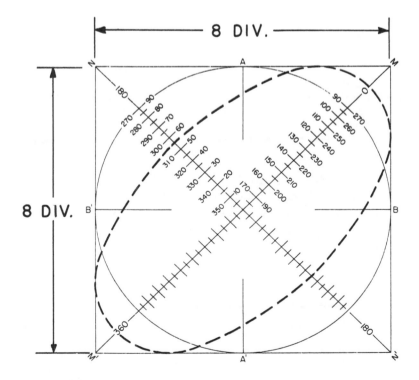

FIG. 4-49. HEWLETT PACKARD WEBB MASK
Reproduced from HEWLETT-PACKARD Application Note No. 29 and
used by permission of HEWLETT-PACKARD COMPANY.

(a)   line   (b)   (x = 2, y = 3) and (x = -4, y = -1)

Review Problems

1.54 $S_h = S_v = 1$ volt/div

The vertical amplifier is set to AC and the horizontal amplifier is direct coupled. The input voltages are

$$v_v = 2 + 2 \sin \omega t$$

$$v_h = 2 + 2 \cos \omega t$$

Assume that ω is high enough so the blocking capacitor does not affect the AC component.

Sketch the trace which will appear on the screen.

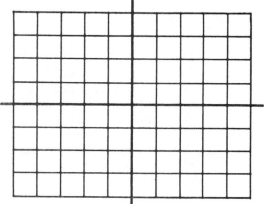

(a)   Channel 1   (b)   the Channel 2 input

2.54   If the electronic switch changes position each time the horizontal sweep is triggered, the scope is in the ALTERNATE dual-trace mode. In this mode, the operation of the electronic switch is synchronized with the sweep waveform as shown below. During the first sweep CH1 is displayed, during the second sweep CH2 is displayed, during the third sweep CH1 is displayed again, etc. On the CH1 and CH2 waveforms shown below, darken the sections which will be displayed on the screen if the scope is set to ALTERNATE dual-trace mode.

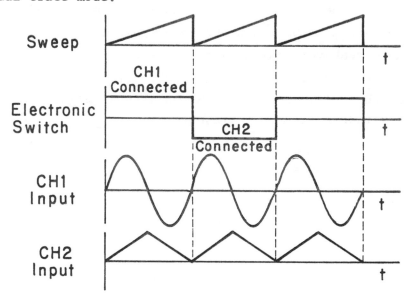

LABORATORY PART III

3.55 In this part you will learn how to check the calibration of the scope and use it to make accurate measurements. You will learn how to operate the remaining scope controls which were not discussed in Parts I and II. The loading effect of the scope on the voltage being measured will be investigated, and the use of the probe will be introduced to reduce this loading effect.

Objectives

When you complete this unit, you should be able to

  (a) check the calibration of the CH1 and CH2 vertical amplifiers
  (b) check the calibration of the time base
  (c) adjust the probe and use it in making measurements
  (d) use the differential amplifier mode to display the difference of two input signals
  (e) measure the common-mode rejection ratio
  (f) apply proper techniques to minimize the effects of hum and noise pickup
  (g) make accurate measurements of voltage and time using the scope

---

  (a) 0 to 90°   (b) 90 to 180°

(Note that these answers are not directly related to the markings on the axis.)

4.50 The mask is calibrated so that the angle in degrees may be read directly without the necessity of making calculations. Before reading the angle, the signal amplitudes and scope gain must be adjusted so that the peak-to-peak vertical deflection is 8 divisions and the peak-to-peak horizontal deflection is the same. If the trace is properly adjusted, what will be seen if the horizontal signal is removed and only the vertical signal is present? _____

_____ (use letters from Fig. 4-49 in your answer).   (a)

If only the horizontal signal is present? _____   (b)

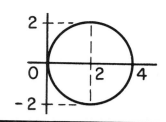

1.55 $S_v = .05$ volts/div, $S_h = 2$ volts/div, the horizontal input is $10 \sin \omega t$ and the vertical input is $.1 + .1 \sin \omega t$. (Note that v_v and v_h are in phase.) Sketch the trace which will appear on the screen if

    (a)   the vertical AC input is used
    (b)   the vertical DC input is used

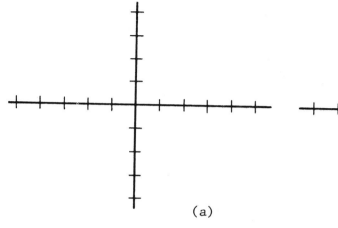

(a)

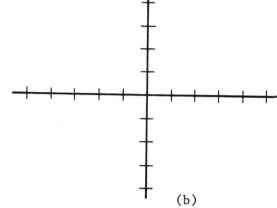

(b)

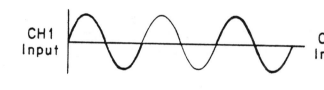

TURN TO FRAME 2.55

Equipment

In addition to your scope you will need the following:

> function generator or sine wave oscillator
> low voltage power supply
> 10X attenuating probe (typical probes are shown in Fig. 3.84, 3.85
>   and 3.86 (pp. 172-176)
> matched pair of precision resistors (each about 500 kilohms)
> two coaxial cable leads, each with a BNC connector to fit your
>   scope on one end and red and black clips on the other end

For frames 3.98 to 3.104 you will also need

> adaptor plug to change 3-prong line plug to 2-prong plug
> 1 kilohm, 10 kilohm, 100 kilohm, and 1 megohm resistors

After taking the necessary precaution, turn on your scope so it will be
warmed up when you are ready to use it.

(a)   line from A to A'   (b)   line from B to B'

4.51   When the trace is properly adjusted, the phase angle may be read directly from the intersection of the ellipse with the scale on the
mask. For the dashed ellipse of Fig. 4-49, θ = _____ or _____   (a)

When the trace is properly adjusted, what relation does the ellipse
bear to all four sides of the outer square (MN'M'N)?

_____   (b)

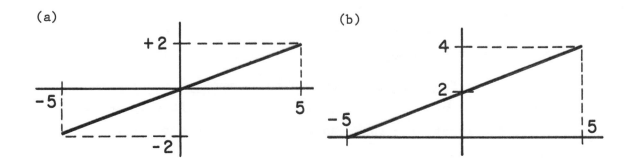

(a)   (b)

1.56 You have now completed Preparation Part I and should understand the types of deflections produced by various signals applied to the vertical and horizontal inputs. Before continuing, go back and review any frames which you missed. Then go to the laboratory and perform Laboratory Part I.

2.55 Suppose the CH1 POSITION control and vertical sensitivity are adjusted so that the CH1 input waveform is displayed only in the top half of the scope screen, and the CH2 controls are adjusted so that the CH2 waveform is displayed only in the bottom half. Then using the ALTERNATE dual-trace mode, the electron beam will alternate between tracing out the CH1 waveform on the top half of the screen and the CH2 waveform on the bottom half. If the sweep is triggered at a high enough frequency, then due to the persistence of the phosphor, <u>BOTH</u> traces will be visible on the scope at the same time.

Assume that the scope is in the ALTERNATE dual-trace mode with the CH1 trace centered in the top half of the screen and the CH2 trace centered in the bottom half. For the waveforms below, sketch the display on the scope screen. Assume S_v = 1 volt/div and the sweep is set to display exactly 1 cycle of the CH1 waveform on the screen.

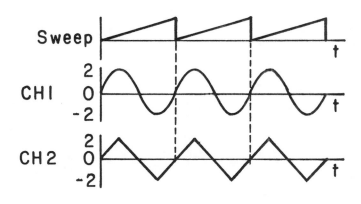

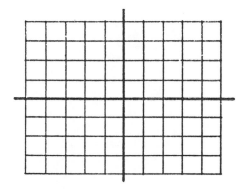

3.57
With no inputs connected to the scope, adjust your scope to obtain a properly focused, centered horizontal line on the screen. (How should you set TRIGGER MODE to trigger the sweep when there is no input?)

Now connect the function generator to the scope and display 10 cycles of a 100 Hz sine wave with the (CH1) vertical sensitivity set to 2 volts/div (calibrated). (How can you tell when the red VAR knob is in the calibrated position?)

922 ONLY: Make sure the X1-X10 knob is also in the calibrated position.

TURN TO FRAME 3.58 ON PAGE 120

(a) 55° or 305°   (b) it is tangent

4.52 | Referring to Fig. 4-52 (p. 118), what adjustments (if any) must be made in each case before reading the phase angle from the Webb mask?

Fig. 4-52a _____

Fig. 4-52b _____

Fig. 4-52c _____

Fig. 4-52d _____

On Fig. 4-49, sketch the trace which should be seen when $\theta = 150°$.   (e)

LABORATORY PART I

1.57 The purpose of this part is to introduce you to the actual operation of the oscilloscope and to demonstrate the relation between voltages applied at the scope terminals and the pattern which will be displayed on the screen. In this part you will put into practice the techniques learned in the Preparation Part I. You must complete this unit before you begin Preparation Part II.

Objectives

Know how to connect, adjust, and use the scope in the X-Y mode of operation (signals applied to both vertical and horizontal inputs) including:

(a) avoid damage to the scope screen.

(b) adjust the position and FOCUS controls.

(c) connect and use the EXT. INPUT on the time base.

(d) operate the vertical sensitivity controls.

(e) use the AC-DC-GND switch.

Have you completed Preparation Part I? _____

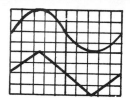

2.56 On the scope screen below, sketch the display if the scope is in the ALTERNATE dual-trace mode, and the CH1 and CH2 traces are centered in the top and bottom half of the screen. Assume S_y = 1 volt/div for both CH1 and CH2, TRIGGER SLOPE is set to -, and TRIGGER LEVEL is set to 0. Waveform (a) below is connected to the CH1 input and the external trigger input and waveform (b) below is connected to the CH2 input. The TRIGGERING SOURCE switch is set to EXTernal and the sweep time/div is set to display exactly one cycle of the CH1 waveform on the screen.

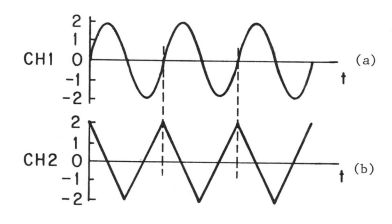

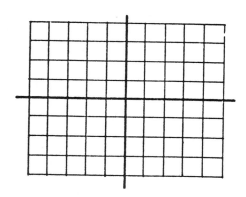

117

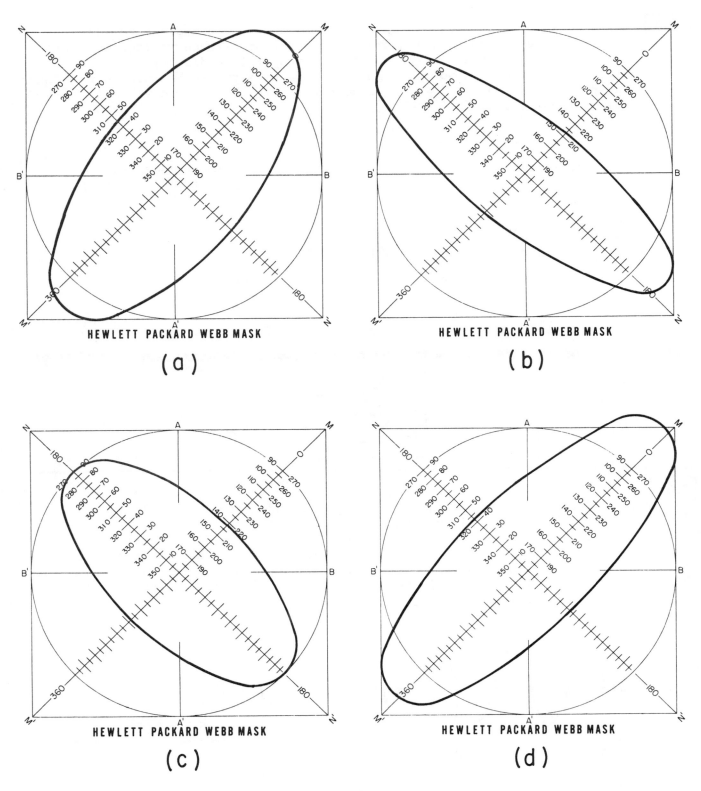

FIG. 4-52
Reproduced from HEWLETT-PACKARD Application Note No. 29 and
used by permission of HEWLETT-PACKARD COMPANY.

If your answer is no, complete the preparation before proceeding.

1.58 Select a position in the lab which is equipped with a suitable* oscillo-scope.

You will also need the following equipment** to provide signals to the scope:

(a) low voltage power supply for DC signals (Tektronix PS503A or equivalent)

(b) digital voltmeter (Tektronix DM501A or equivalent) to set DC power supply voltage

(c) function generator for sine, triangular and square waves (Tektronix FG501 or equivalent)

(d) an oscilloscope lead kit

If the above equipment is not already at your position, obtain it from your lab instructor or supervisor. He will demonstrate the proper way to operate this equipment. Also obtain a continuity tester and an R-C network from the supervisor. Other necessary leads should be available in the laboratory.

---

*See Table B-1, p. 246, for a list of suitable oscilloscopes.
**See Appendix A for specifications for the required auxiliary equipment.

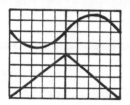

2.57 With the scope in the dual-trace mode, if the TRIGGER SOURCE is set to INTernal, either the CH1 or CH2 input may be selected as the triggering signal. That is, either the CH1 or CH2 input is connected internally to the trigger pulse generator, and the other channel input does not affect the time at which triggering occurs.

For the previous frame, if the CH1 input is disconnected from the external trigger input and the trigger source is set to internal, which vertical input must be selected as the triggering signal in order to have the scope

display shown in the above answer? _____ (a)

If CH2 is now selected as
the trigger signal, sketch
the resulting scope display   (b)
(no change in TRIGGER SLOPE
and LEVEL controls).

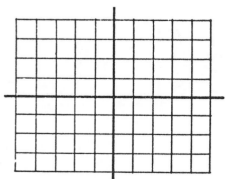

Sweep Magnifier

3.58

922
ONLY

Any portion of the trace can be expanded horizontally up to 10 times by rotating the X1-X10 knob clockwise. To expand any portion of the trace, move that portion to the center of the screen and rotate the X1-X10 knob clockwise until the trace is expanded as much as desired. Adjustment of the horizontal position may be necessary. Try this.

When the X1-X10 knob has been rotated fully clockwise until it is past the click-stop, the trace has been expanded horizontally by a factor of

_____.   (a)

CAL
X-Y
ONLY

The trace can be expanded horizontally by a factor of 10 (or 5). To expand a portion of the trace move that portion to the center of the screen and enable the X10 (X5) control.* Further adjustment of the horizontal position may be needed. Try this.

When the X10 (X5) control has been enabled, the trace has been expanded

horizontally by a factor of _____.   (a)

When in the X10 (X5) position, to determine the true value of time/div

(multiply/divide) _____ the setting of the   (b)

SEC/DIV switch by _____.   (c)

_____

*2213: Make sure that the red knob in the center of SEC/DIV is in the CAL position, and then pull it out.

(a) increase horiz. gain   (b) increase vert. gain
(c) increase both vert. and horiz. gain   (d) adjust vert. position
(e) Your ellipse should have its major axis on N-N'; it should be tangent to all four sides of the square; it should intersect the scale at 150°.

4.53

922
ONLY

In the given network, v_1 and v_2 do **not** have a common ground terminal. The phase angle of v_2 with respect to v_1 is to be measured by the Webb mask. Indicate the proper connections to the scope.

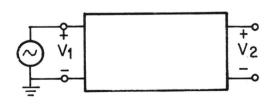

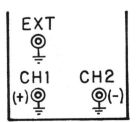

Familiarization with the Scope

1.59 This text is designed for use with two different types of oscilloscopes. One of these types has calibrated range switches for both the X (horizontal) and Y (vertical) inputs. We will refer to this type of scope as "CAL X-Y". The other type of scope, which is similar to the Tektronix T922, has a calibrated range switch only for the vertical input. We will refer to this type of scope as "922". Although most parts of the text apply to both types of scopes, some frames apply to "CAL X-Y ONLY" and some to the 922 (or similar scopes) only. If you do not have a 922 scope, refer to the table on page 246 or ask your lab instructor which type of scope you have. If you have a "CAL X-Y" scope, omit those parts of the text marked "922 ONLY". If you have a 922 or similar scope, omit those parts marked "CAL X-Y ONLY".

For your scope, the parts marked _____ should be omitted. (a)

Most oscilloscopes use BNC coaxial cable connectors of the type shown to the right. How many BNC connectors does your scope have? _____ (b)

(a) CH1 (b)

2.58 Sketch the display on the scope screen below if TRIGGER LEVEL is set to 0, TRIGGER SOURCE is set to INT (CH1), TRIGGER SLOPE is set to +, and $S_v = 1$ volt/div for both CH1 and CH2. Waveform (a) is the CH1 input and waveform (b) is the CH2 input. The scope is using the ALTERNATE dual-trace mode and sweep time/div is set to display exactly one cycle of the CH1 waveform. Assume that the CH1 and CH2 traces have been centered in the appropriate halves of the scope screen and that the electronic switch is in the CH1 position immediately after t = 0.

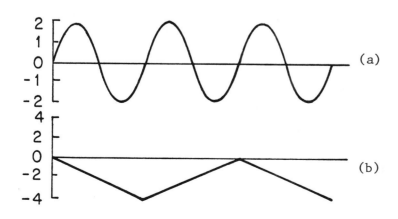

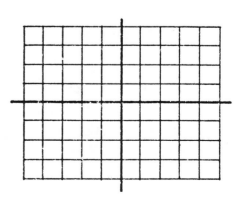

(a)   10 (5 for X5)

(b)   divide (since the spot moves faster, there is less time per division)

(c)   10 (5 for X5)

3.59   In the X10 (X5) position, the fastest possible sweep rate is

_____ microseconds/div.

---

v_1^+ to external horizontal (X) input

v_2^+ to +INPUT(CH1), v_2^- to −INPUT(CH2);   oscillator ground to scope ground

---

Lissajous Figures

4.54   Lissajous figures can be used to determine the ratio between two frequencies.   The ellipse which you obtained when making phase shift measurements
is an example of a Lissajous figure with a one-to-one (1:1) frequency ratio.
In general, a Lissajous figure can be obtained by applying two sinusoidal
signals of different frequencies to the horizontal and vertical inputs of
an oscilloscope.   In order to obtain a stable Lissajous pattern on the
scope screen, the two frequencies must be in the ratio of small whole numbers such as 3:2, 5:6, etc.

If V_v = 2.03 sin 200πt and V_h = 1.27 sin 300πt, would a stable Lissajous
figure be obtained? _____

(a)   922 ONLY   (if you have a CAL X-Y scope)
      CAL X-Y ONLY   (if you have a 922 scope)
(b)   3 (for 922 and most other scopes)

1.60   Turn to frame 1.61 unless you have a Tektronix T922 scope.

922
ONLY

The Tektronix T922 oscilloscope has one
horizontal amplifier (as part of the
TIME BASE) and two vertical amplifiers.
The controls for the horizontal amplifier
are located in the TIME BASE section of
the scope. On your lab scope, locate the
Time Base, Vertical Amplifier Channel 1
and Vertical Amplifier Channel 2 sections.
Also identify the horizontal input
connector and the CH1 and CH2 input
connectors. On your scope, the horizontal
input connector is labeled (X, Y, AMPLIFIER)

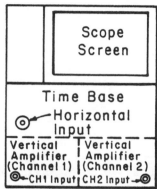

_____,   (a)

The CH1 input connector is labeled _____,   (b)

and the CH2 input connector is labeled _____. (c)

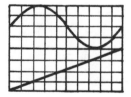

2.59   In the CHOP dual-trace mode, the electronic switch switches rapidly between
the CH1 and CH2 inputs and is not synchronized with either the CH1 or CH2
input or the sweep. On the CH1 and CH2 waveforms below, darken the sections which will be displayed on the scope screen if the scope is set to
the CHOP dual-trace mode.

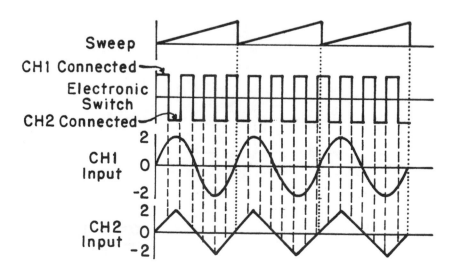

| 3.60 | With the control in the X10 (X5) position and the SEC/DIV set to 1 ms, the following trace is observed:

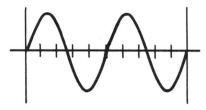

What is the frequency of the signal applied to the vertical input?

_____

After writing your answer, verify it by observation.   Then reset your scope for X1 calibrated operation.

Yes (Since the frequency ratio is $200\pi/300\pi = 2:3$)

| 4.55 | Suppose the x signal of Fig. 4-55a is applied to the horizontal input of an oscilloscope and the y signal (Fig. 4-55b) is applied to the vertical input.   These two signals differ in frequency by a factor of two.   The scope trace can be graphically constructed by examining x(t) and y(t). At point ①, x(t) is at a maximum and y(t)=0.   During the time that y(t) completes two full cycles (point ① to ⑤), x(t) will vary from a positive maximum (①) through zero to a negative maximum (③), back through zero to a positive maximum (⑤) tracing a "figure eight" on the oscilloscope screen.   Plot the points on the Lissajous figure which correspond to the points ⓐ and ⓑ on the waveforms.

$x = A \cos \omega t$

$y = B \sin 2\omega t$

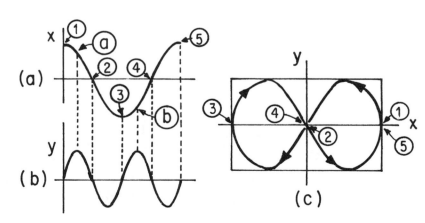

FIG. 4-55.   GRAPHICAL CONSTRUCTION OF A
LISSAJOUS FIGURE

(a) X   (b) Y   (c) AMPLIFIER

Skip this frame if you have a T922 scope.

1.61

CAL
X-Y
ONLY

Most dual-trace oscilloscopes have three main groups of controls on the
front panel, consisting of two identical vertical sections and one horizontal section. Each vertical section includes a VOLTS/DIV switch, a
position control, an AC-DC-GND switch, and a BNC input connector. The
vertical sections are labeled CH1 and CH2 or A and B. Locate these
sections of your scope front panel. The horizontal section normally includes a TIME/DIV(SEC/DIV) knob, a position control, an external trigger
input connector, and various triggering controls. Locate the horizontal
section on your scope.

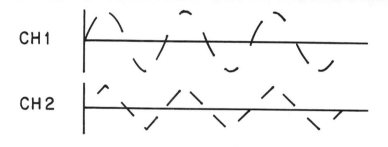

2.60

Notice from the last frame that each time the scope is triggered, different
sections of the CH1 and CH2 waveforms will be traced on the screen. If
the scope is triggered at a high enough frequency, then the CH1 traces will
overlap and the scope will display the full CH1 waveform with no visible
chopped lines. Similarly, the CH2 display will not have visible chopped
lines. Sketch the CH1 and CH2 waveforms from frame 2.59, as they will
appear on the scope screen. (Note that the electronic switch is not
synchronized with the waveforms.) S_v = 1 volt/div.

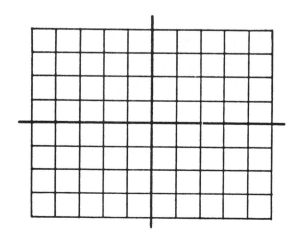

2000 Hz (1000 Hz for X5)

In the X10 (X5) position, the sweep rate is 0.1 ms/div (0.2 ms/div)

The period of the wave is 5 div x .1 ms/div = .5 ms (1 ms for X5), so

f = 2000 Hz (1000 Hz).

Checking the Calibration of the Vertical Amplifier Gains

3.61 If the scope is to be used to make accurate voltage measurements, the vertical amplifiers must be properly calibrated. To check the calibration, we will use a power supply to produce a DC voltage which we will measure with both the oscilloscope and an accurate digital voltmeter. If the values of the DC voltage measured with the oscilloscope and with the digital voltmeter differ significantly (more than 3 to 5%, depending on the type of scope), then the vertical amplifier (is/is not) _____ properly calibrated.

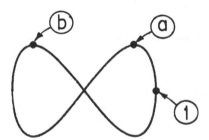

4.56 The frequency ratio can be determined by examining the Lissajous figure. In Fig. 4-55c the figure has two vertical positive maximums (at points (a) and (b)), and one horizontal maximum (point (1)). In the time it takes to traverse the figure, y goes through twice as many maximums as x. Therefore the ratio of the frequency of y(t) to that of x(t) is

$$\frac{f_y}{f_x} = \underline{\hspace{4cm}}$$

1.62 The scope has a 3-wire line cord. Two of the wires supply AC power to the scope and the third wire is a safety ground. The third wire is connected directly to the metal scope case or front panel and to the scope chassis. When the scope is plugged in, these metallic parts are connected to the building ground through the 3-wire line cord. When the scope is plugged in, is it possible to have a high voltage between the scope case or front panel and the building ground?

_____ Explain.   (a)

_____   (b)

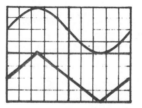

2.61 Generally, ALTERNATE is the best dual-trace mode to use with "fast" sweep rates (about .5 msec/div or faster). With slower sweep rates, the scope display can be seen to alternate between tracing out the CH1 and CH2 wave-forms and both waveforms may not be visible all the time. For "slow" sweep rates (about 1 msec/div or slower), CHOP is usually the best dual-trace mode to use. The sweep rate will still be fast enough for the CH1 and CH2 waveforms to appear without chopped lines.

Which dual-trace mode should be used for the following sweep time/div settings?

(a) 2msec/div _____

(b) 5μsec/div _____

is not

| 3.62 | In order to read the scope display accurately, a sharply defined trace
is necessary. Set the scope to display a horizontal line when the inputs
are grounded.

Try to focus the trace as sharply as possible for several different
settings of the intensity control. With the focus control set to an
optimum position, what is the relation between the intensity and the
thickness of the trace? _____

_____

Set the focus and intensity controls to what you think is an optimum
position for making accurate measurements with the scope.

Now turn down the intensity slightly. If the trace is still clearly
visible but you can focus for a finer, sharper trace, your original
settings were probably not optimum.

2/1   (or 2:1)

| 4.57 | As the phase angle between the two input waveforms changes, the shape of
the Lissajous figure will change from "fully open" as in Fig. 4-55 (c) to
"fully closed" as in Fig. 4-57. Recall that as the phase angle varied from
$0°$ to $90°$ in the ellipse method (a Lissajous figure for frequency ratio
1 : 1), the ellipse varied from a straight line to a full circle. The
circle corresponds to the "fully open" figure and the line to a "fully
closed" figure. Determining the frequency ratio of a closed Lissajous figure is difficult and <u>measurements should always be made on a fully open
figure if possible</u>.

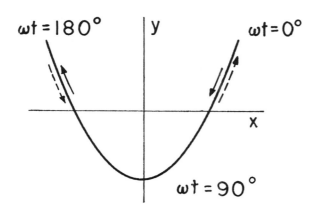

$x = A \cos \omega t$
$y = B \cos 2\omega t$

The trace of a fully _____ (a)
Lissajous figure has definite
endpoints; but the trace of a
fully _____ figure   (b)
is a continuous smooth curve.

FIG. 4-57. A FULLY CLOSED LISSAJOUS FIGURE

128

(a)   no   (b)   The case or front panel is connected to the building
                      ground through the line cord.

1.63   Examine your scope and locate the CH1 and CH2 vertical input terminals.
       The figure below shows typical input terminal locations. Using the
       continuity tester, determine if the round grounding pin on the line plug
       is connected directly to each of the following (check if there is a
       connection):

       ☐ outer conductor on CH1 input connector
       ☐ outer conductor on CH2 input connector
       ☐ outer conductor on other BNC connectors
       ☐ grounding jack, marked ⏚ or GND (if your scope has one)

       Are all of the above terminals connected together? _____   (a)
       Are all of these terminals connected to the building ground when the
       scope is plugged in? _____   (b)

EXT (Labeled X on T922)

— INNER CONDUCTOR —

— OUTER CONDUCTOR —
(Ground)

CH1   Grounding   CH2
             Jack

TYPICAL INPUT TERMINAL LOCATION

(a)   CHOP   (b)   ALTERNATE

2.62   This completes Preparation Part II. You should now understand how to dis-

       play waveforms on the scope as a function of time. You should understand

       the functions of the TRIGGER SLOPE, LEVEL, SOURCE, and MODE controls on

       the TIME BASE. You should also know the difference between the ALTERNATE

       and CHOPPED dual-trace modes.

       Now go to the lab and do Laboratory Part II.

the greater the intensity, the thicker the trace

3.63 To check the CH1 vertical amplifier calibration, set the scope to display
a horizontal line on the bottom line of the screen when the CH1 input
is grounded. Then, using the digital voltmeter, set the power supply
to exactly 6 volts. If a deflection of 6 divisions is desired, the

sensitivity switch should be set to _____ VOLTS/DIV with the (a)

red VAR knob in the _____ position. (b)

After writing your answer, verify it by observation. The observed
deflection may not be exactly 6 divisions, so calculate the percent error.

Percent error = $\dfrac{|\text{Actual oscilloscope deflection } - \text{ expected deflection}|}{\text{Expected deflection}} \times 100\%$

Percent error for your scope = _____. (c)

(a) closed   (b) open

4.58 Both of the Lissajous figures given below represent the same frequency
ratio.
Should the left or right one be used to determine this ratio? _____ (a)
This figure has <u>three</u> positive vertical maximums and _____ (b)
positive horizontal maximums.
For a fully open Lissajous figure, the ratio of the frequency driving
the y-axis to that driving the x-axis is given by

$$\frac{f_y}{f_x} = \frac{\text{no. of positive vertical maximums}}{\text{no. of positive horizontal maximums}}$$

From the Lissajous figure chosen in (a) above, the frequency ratio is
$f_y/f_x =$ _____. (c)

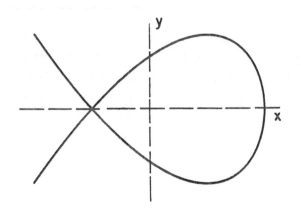
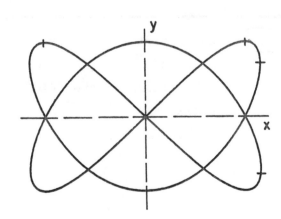

(a)   yes   (b)   yes

1.64   The line cord ground is <u>not</u> a reliable ground and should not be used
to carry signals being observed on the scope. That is, always connect
a separate ground wire directly to the signal source. The oscillator
is being used to supply a signal to the scope in the diagram below.
Assume that one side of the oscillator output is grounded to the
building ground through its line cord.

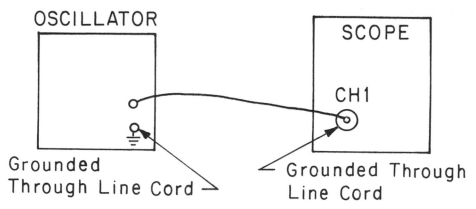

Draw a line to indicate what other connection (if any) should be made   (a)
in the above circuit. Would this connection be necessary if the line
cord grounds were perfect? _____   (b)

LABORATORY PART II

2.63   In Lab Part I you operated the scope only with an external horizontal input.
In this part you will use the TIME BASE to generate a horizontal sweep signal. You will learn how to display periodic waveforms on the screen, and
you will also learn how to adjust the triggering controls.

<u>Objectives</u>

1.   Be able to select the appropriate settings for the following triggering
     controls: TRIGGER SOURCE, TRIGGER MODE, TRIGGER SLOPE, and TRIGGER
     LEVEL.

2.   Use the SEC/DIV knob to select the proper sweep rate.

3.   Display a periodic waveform selecting all necessary settings and inputs.

4.   Display two waveforms in the dual-trace mode.

<u>Equipment</u>

In addition to the scope and scope cables, you will need the following:

    (a)   low voltage DC power supply
    (b)   function generator
    (c)   sine-wave oscillator (or a second function generator)

Have you completed Preparation Part II? _____

(a)   1 volt/div   (b)   calibrated

(c)   If the error is greater than 4%, then the CH1 vertical amplifier gain
      on your scope may need adjustment. If such is the case, ask your lab
      supervisor to verify your measurements.

3.64   Calibration could also have been done by setting the gain to 2 volts/div
        and adjusting for 3 squares deflection. Why would this be less desirable

        than using 6 squares deflection as above? _____

        _____

(a)   right   (b)   2   (c)   3/2

4.59   Determine the frequency ratio for each of the figures given below.

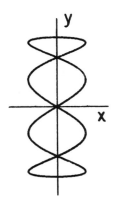

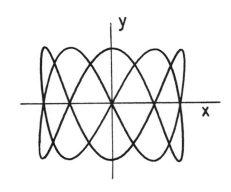

(a) f_y/f_x = _____   (b) f_y/f_x = _____

(a) connect a wire from the scope ground to the oscillator ground (since the line cord grounds are not reliable)

(b) NO

1.65 Burned spots appear on the scope screen as small, dark areas. Does

your scope screen have any burned spots? _____   (a)

If your scope doesn't have any burned spots, ask your lab instructor

to show you a CRT which does. This damage to the CRT was caused by

having the intensity set too _____ when the spot was   (b)

_____.   (c)

If your answer is NO, go back and complete the preparation before proceeding.

Triggering the Sweep

2.64 (a) What is the source of the triggering signal when the TRIGGER SOURCE
is set to INTernal? _____

to LINE? _____

to EXTernal? _____

(b) Explain the difference between AC and DC TRIGGER COUPLING.

_____

_____

(c) What is the effect of the TRIGGER SLOPE control?

_____

_____

(d) What is the effect of the TRIGGER LEVEL control?

_____

_____

The accuracy would be less. (The percent error in reading the scope
trace would be larger even though the absolute error would be about
the same.)

3.65   If the gain adjustment has been properly set, the vertical amplifier
calibration should be correct for all settings of the VOLTS/DIV switch.

Check the calibration of the vertical amplifier on the 2 volts/div range.

DC power supply voltage used: _____ Percent error: _____ (a,b)
Check the calibration of the vertical amplifier on the .5 volt/div range.

DC power supply voltage used: _____ Percent error: _____ (c,d)

(a) 1/4   (b) 5/2

4.60   Summary: Three basic methods for measuring the phase angle of a
voltage v_2 with respect to a reference voltage v_1 have been studied.

1. The <u>dual-trace</u> method consists of two steps:

   a. Using v_1 as a CH1 vertical input, display v_1 as a function of
       time and calibrate the time axis for 20°/division (or some other
       convenient value).

   b. Using the dual trace mode (with CH1 as trigger source) display
       v_2 and observe the phase difference.

2. The <u>triggered sweep</u> method consists of two steps:

   a. Using v_1 as an external trigger input and as the vertical input,
       display v_1 as a function of time and calibrate the time axis in
       degrees.

   b. Without disturbing the trigger circuit, change the vertical input
       to v_2 and observe the phase difference.

3. The <u>ellipse</u> method consists of the following steps:

   a. Display v_2 vs v_1 on the screen.

   b. Compute the phase angle from the ratio of distances measured on
       the resulting ellipse. (The angle may also be read directly by
       using a Webb mask.)

(b)   high   (c)   stationary (not moving)

1.66   As a precaution against burning the screen, always turn the <u>intensity</u>
control all the way down (fully counterclockwise) <u>before</u> turning the
scope on.

Take the necessary precaution against burning the screen, and then
turn the scope ON so that it will be warmed up by the time you are
ready to use it.

Also turn on the function generator and DC power supply so they
will be warmed up when you need them.   The screen is ruled off into
divisions one centimeter square.   The ruled portion of the entire
screen is _____ cm by _____ cm.   Each division is marked off   (a)
into subdivisions along the axis.   How many subdivisions
are contained in each division? _____   (b)

(a)   from the vertical amplifier output; from the 60 Hz AC line; from the
       external trigger input
(b)   On AC, trigger circuit responds to AC component of triggering signal
       only; on DC, trigger circuit responds to entire triggering signal.
(c)   determines whether triggering occurs on positive or negative slope of
       triggering signal
(d)   determines the voltage level at which triggering occurs

---

2.65   Examine the TRIGGERING MODE switch on your scope. Note that in addition
922   to the AUTO and NORMal positions, the switch also has a TV position.
and   This position is for triggering the scope on television signals. The TV
2213   position will not be discussed further or used in this book.

Other   Examine your scope to determine how AUTO and NORMal triggering is selected.
Scopes   On some scopes, AUTO is selected by using the TRIGGER LEVEL control.

ALL   What triggering mode should be used if we want the sweep to trigger when
         no signal is present? _____
         Select this triggering mode on your scope.

(a) 12 to 16 volts should have been used (a smaller voltage would be less accurate)

(c) 3 to 4 volts

(b,d) If error > 4% for either switch setting, there may be something wrong with your scope. Ask the supervisor to check.

3.66   In frame 3.63 you checked the calibration of the CH1 vertical amplifier. Repeat frame 3.63 for the CH2 vertical amplifier and if the gain is not properly adjusted call your lab supervisor.

4.61   You have now completed Preparation Part IV.

You should know how to measure phase shift by the dual-trace method, the triggered sweep method and the ellipse method. You should also know how to determine the ratio of two frequencies using Lissajous figures. Now go to the lab and do Lab Part IV.

1.67

922 ONLY

Examine the block diagram of the scope (Fig. 1-67). The input to the horizontal amplifier may come from either the _____ (a) or the _____.

In this part we will use an external horizontal input; the use of the sweep voltage from the TIME BASE will be explained in Part II.

The switch on the T922 oscilloscope which connects the horizontal amplifier to an external horizontal input is the 5-position lever switch marked TRIGGERING SOURCE located in the TIME BASE section of the oscilloscope. In the top 4 positions, this switch connects the TIME BASE to the horizontal amplifier. In its lowest position, this switch connects the external horizontal input to the horizontal amplifier. Set the SOURCE switch on your scope so that the horizontal input may come from an external source.

What is this switch position labeled? _____ (b)

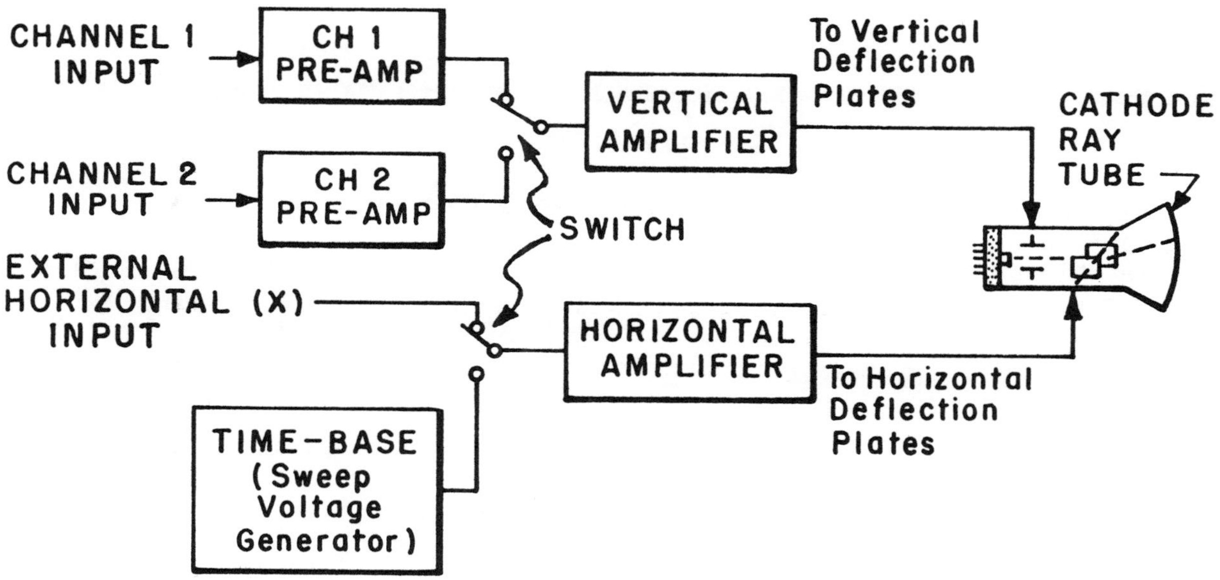

FIG. 1-67. SIMPLIFIED BLOCK DIAGRAM OF SCOPE

You should have selected the AUTO mode.

2.66   What precaution should be taken before turning on the scope?

_____ (a)

Take this precaution and explain the reason for it.

_____

_____ (b)

3.67   When trying to make accurate voltage measurements with the scope, we should always use as large a deflection as possible because

_____ (a)

_____

Connect a sine-wave oscillator to the scope and set the output amplitude to about the middle of its range and observe the sine wave.

Leaving the red knob in the CALIBRATED position, rotate the VOLTS/DIV switch until the waveform <u>just</u> goes off scale (peak-to-peak deflection greater than 8 divisions); then back it down <u>one</u> position. The deflection should now be the maximum possible for the given input voltage.

       observed peak-to-peak deflection _____ div.   (b)

For three different input voltage amplitudes (different enough so you have to change ranges on the scope), record the maximum obtainable peak-to-peak deflection (VOLTS/DIV still calibrated):

_____ div.   _____ div.   _____ div.   (c)

LABORATORY PART IV

4.62   In this part you will measure the phase shift of a network at different frequencies using first the dual-trace method and then the ellipse method. The results obtained by the two methods should agree closely if you work carefully.

<u>Objectives</u>

1. Measure the phase difference between two sinusoidal voltages using the dual-trace method, the triggered sweep method, the ellipse method, and the Webb mask. Determine the sign of the phase angle when possible.

2. In each case, adjust the scope to obtain maximum accuracy in your measurements.

3. Measure the frequency ratio of two sine waves using Lissajous figures.

<u>Equipment</u>

In addition to an oscilloscope, you will need the following:

     active phase shift network; Webb mask (optional)
     sine wave oscillator or function generator

Record the identification numbers on the back of the phase shift network and on the oscillator in Table 4-1 (p. 226). Use the same network and oscillator throughout Lab Part IV. You will use Table 4-1 to keep a record of phase shift measurements made by different methods so that you can easily check your answers.

Turn on the equipment so it will be warmed up when you are ready to use it.

138

1.68 Examine the block diagram of the scope (Fig. 1-68). The input to
the horizontal amplifier may come from either the _____

CAL   or the _____.   (a)

X-Y   In this part, we will use an external horizontal input; the use of

ONLY the sweep voltage from the TIME BASE will be explained in Part 2.

The switches in the diagram are shown set in the position for the
X-Y mode of operation. In this mode, the external horizontal input
signal should be applied to the (CH1/CH2) _____ input in order   (b)
to cause a horizontal (X) deflection, and a signal should be applied
to the _____ input to cause a vertical (Y) deflection.   (c)

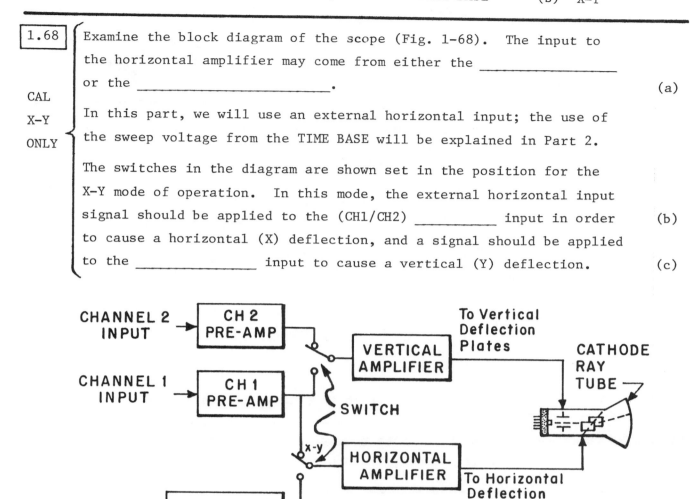

FIG. 1-68. SIMPLIFIED BLOCK DIAGRAM OF A TYPICAL CAL X-Y SCOPE

You should have turned down the intensity control because a high intensity
spot would burn the screen.

TURN TO THE NEXT PAGE

(a) the per cent error in reading the scope will be smaller when the deflection is large

(b,c) All of your answers should be in the range from 4 div. to 8 div.
peak-to-peak regardless of the input voltage. If any of your answers are not in this range, go back and try three more different
input voltages.

3.68 | Figs. 3-68x and y show the same sine wave displayed for two different
settings of the position controls. In which case can you read the
peak-to-peak deflection most accurately? _____ (a)

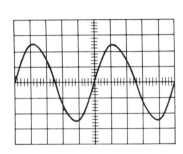

FIG. 3-68x   Fig. 3-68y

Adjust the oscillator output voltage to 3.25 volts peak-to-peak. Do
this as accurately as you can.

peak-to-peak deflection _____ div. (b)

Did you turn down the intensity before you turned the scope on?
If you didn't, turn it down quickly before it is too late.

Phase Measurement by the Dual Trace Method

4.63 | In order to obtain accurate results when using the dual trace method,
the time axis of the scope must be carefully calibrated. For each pair of
traces below, check the one which should give the most accurate results.

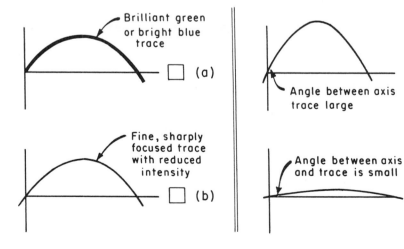

(a)   TIME BASE or CH1 PREAMP output   (b)   CH1   (c)   CH2

1.69

922
ONLY

As shown in the figure below, there are three or four* rectangular buttons located between the two sets of vertical amplifier controls. These buttons are called the vertical mode buttons. Locate the vertical mode buttons on your oscilloscope. These buttons determine how the oscilloscope displays the outputs from its Channel 1 and Channel 2 vertical amplifiers. Until further instructed, we will be using only the Channel 1 vertical amplifier, so depress the CH1 vertical mode button.

Vertical Mode
Switches*

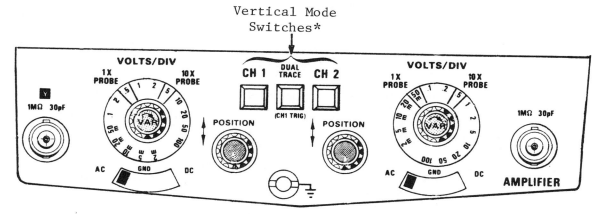

(Used by permission of Tektronix, Inc.)

_____

*The Tektronix T922 has an optional 4th vertical mode button labeled "DIFF".

Note: Most of the material in the following frames applies to all types of oscilloscopes. However, some parts are labeled "922", "CAL X-Y" or "2213".

If you have a Tektronix 922 or similar oscilloscope, work the parts labeled "922" and omit the parts labeled "CAL X-Y" or "2213".

If you have a Tektronix 2213 oscilloscope, work the parts labeled "CAL X-Y" and the parts labeled "2213"; omit the parts labeled "922".

If you have another type of calibrated X-Y oscilloscope, work the parts labeled "CAL X-Y" and omit the parts labeled "2213" or "922".

(a) Fig. 3.68y   (b) 6.5 div. (for best accuracy you should have ad-
                        justed the position controls like for
                        Fig. 3.68y)

3.69   Leave the oscillator set as in the preceding
frame. Connect the DC power supply in series
with the oscillator as shown. Set the DC
supply to 20 volts and observe v_1 using the
DC scope input. (If you have trouble getting
the sweep to trigger, use AUTO trigger.)
Could you measure the AC component of v_1
accurately when the scope is set to DC?

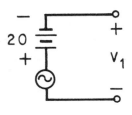

_____ Explain.

_____   (a)

Set the scope to measure the AC component of v_1 as accurately as
possible.

        observed peak-to-peak deflection _____   (b)

The DC component of a waveform can be determined by observing the
amount of shift when the scope input is switched from AC to DC.
Measure the DC component of v_1 as accurately as you can by this
method.

   number of divisions shift _____   (c)

   DC component of v_1 _____ (include the sign)   (d)

(b) is more accurate because a fine, sharply focused trace is easier to
     read

(c) is more accurate because the exact point at which the trace crosses
     the axis is easier to determine if the angle of crossing is large

---

4.64   For Fig. 4-64x, each division represents _____ degrees. (a)

For Fig. 4-64y, each division represents _____ degrees. (b)

In which case could the phase shift be determined more accurately? ____ (c)

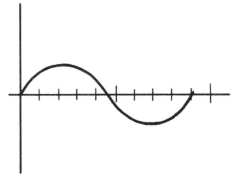

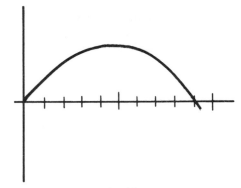

FIG. 4-64x   FIG. 4-64y

142

TIME/DIV.

←Time Base is
Connected in
these Positions

X-Y

1.70

CAL
X-Y
ONLY

The physical location of the
switch(es) which select X-Y
mode of operation depends on
the type of scope being used.
One common location is on the
TIME/DIV (SEC/DIV) switch as
shown. If the TIME/DIV (SEC/
DIV) knob on your scope has an
X-Y position, select that position.

If your scope has a mode button labeled "X-Y", select this mode.

If the triggering source switch has an X-Y position, also select this
position.

2.67

Throughout Laboratory Part II, we will use the sweep mode of operation
rather than the X-Y mode. We will be using the channel 1 vertical input
only.

Follow these steps to obtain a centered horizontal line on your scope
screen:

(a) Turn on the scope and allow it to warm up.

(b) Select the CH1 vertical mode.

(c) Ground the CH1 and CH2 inputs using the AC-DC-GND switches。

(d) Set the TIME/DIV (SEC/DIV) switch to 1 millisecond/division (cali-
     brated).

(e) If your scope has a delayed sweep capability, turn the delayed
     sweep off.

     2213: Set HORIZ MODE to NO DLY. Also set TRIGGER VAR HOLDOFF
            to NORM.

(f) Set the trigger SOURCE switch to INTernal.

     2213: Set the INT source selector to VERT MODE.

(g) Make sure the controls are set for AUTO triggering.

(h) Adjust intensity, position and focus controls as required to obtain
     a sharp, centered horizontal line.

(a)   No. The peak-to-peak deflection is too small.

(b)   6.5 div

(c)   about 4 div (with S_v = 5 volts/div) (If you measured 2 div or less, go back and try for a better accuracy. Use the position control if necessary.)

(d)   Any answer in the range -18 to -22 is acceptable. It would be surprising if you measured exactly -20 volts since there are errors inherent in the meter as well as in the scope.

3.70 Using the power supply and digital voltmeter, adjust the (CH1) VOLTS/DIV and associated VARiable knob so that the vertical sensitivity is 2.5 volts/div. Do this as accurately as you can.

> power supply voltage used _____ (a)
>
> volts/div setting   _____ (b)
>
> deflection   _____ (c)

(a)   40°   (b)   20°   (c)   Fig. 4-64y

4.65 When the sweep is triggered, the exact starting point of the trace may be hard to see, especially at high sweep rates.

Which of the two traces below will provide more accurate results when calibrating the time axis? _____ (especially at high sweep rates)

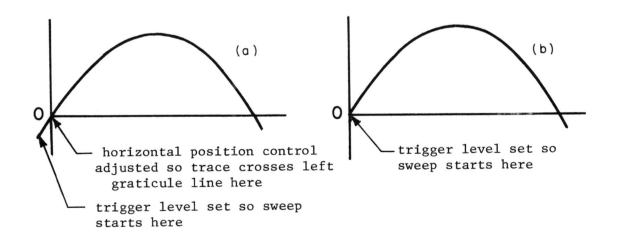

1.71 The AC-DC-GND switch on each vertical pre-amp can be used to select the AC input, the DC input, or to ground the input to the vertical pre-amp.

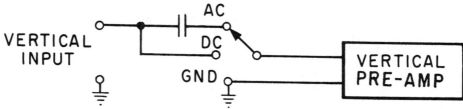

When set to AC, there is a blocking capacitor connected between the

_____ and the _____ (a,b)

When set to GND, the input to the vertical pre-amp is

_____ volts.   (c)

2.68

922
ONLY

The sweep rate is adjustable by means of the grey knob labeled SEC/DIV and the red X1◄►X10 knob in its center. The calibrated sweep times imprinted on the SEC/DIV dial apply only when the red knob is in the X1 position (rotated fully counterclockwise). For this reason, the X1 position is often called the calibrated position.

Try operating the red X1◄►X10 knob and note that it has click-stop positions when rotated fully counterclockwise (X1 or calibrated position) and when rotated fully clockwise (X10 position).

On SEC/DIV, how can you tell when the X1◄►X10 knob is in the calibrated position?_____

(a) 20 volts   (b) 2 volts/div   (c) 8 div

(10 volts and 4 div could have been used, but this would be less accurate)

Checking the DC Balance

3.71 If the DC balance of your scope is not properly adjusted, the vertical preamplifier will have a small DC output voltage even when the vertical input is grounded. To check the DC balance for channel 1 of your scope, proceed as follows:

(1) Ground the CH1 input, select CH1, set VOLTS/DIV to its maximum calibrated value, and display a centered horizontal line on the screen.

(2) Observe the horizontal line as you rotate VOLTS/DIV to its most sensitive position. If the maximum deflection of the line exceeds 0.2 divisions from the center, the DC balance on your scope needs adjustment.

Carry out the above steps. Maximum observed deflection of horizontal line?

_____ divisions.   (a)

Does the CH1 balance on your scope need adjustment? _____   (b)

(a) is more accurate because it is easy to see the exact point at which the trace crosses the axis, but it is difficult to see the exact point at which the trace starts

---

4.66 We will use a phase shift network of the following form*:

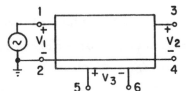

Note: Terminals 5 and 6 are not directly connected to terminals 2 and 4.

v_1 and v_2 (do/do not) _____ have a common ground. Therefore, (a)
to measure the phase angle between v_1 and v_2, the dual-trace method
(can/cannot) _____ be used. The dual-trace method   (b)
(can/cannot) _____ be used to measure the phase angle   (c)
between v_3 and v_1 because _____.

---

*See Fig. A-1 on page 245 for the complete circuit diagram.

(a) vertical input   (b) vertical pre-amplifier   (c) 0

1.72 Set up the scope as follows:

1) Temporarily ground the CH1 and CH2 inputs with the AC-DC-GND switches.

2) Turn the CH1 and CH2 VOLTS/DIV (sensitivity) knobs fully counter-clockwise ↶ . (largest volts/div).

922
ONLY { 3) Rotate the red X1 X10 knob on the SEC/DIV switch fully counter-clockwise.

4) Turn up the INTENSITY control and adjust the POSITION controls to obtain a centered spot on the screen. The horizontal position control is located near the SEC/DIV knob. The vertical position of the spot is changed by the (CH1/CH2) _____ position control.

   WARNING: WHEN WORKING WITH A STATIONARY SPOT, USE A VERY LOW
             LEVEL OF INTENSITY.

5) Adjust the INTENSITY so that the spot is clearly visible, but it is not any brighter than necessary. If the screen glows around the spot, it is too bright.

The X1 X10 knob is rotated fully counterclockwise and is in the click-stop.

2.69 The sweep rate is adjustable by means of the SEC/DIV knob and the red VARiable knob* usually located in its center. The calibrated sweep times printed on the SEC/DIV dial apply only when the red VARiable knob is in the calibrated position (rotated fully clockwise). Rotating the VARiable knob counterclockwise increases the sweep time/division by as much as 2.5 times. For example, .1 sec/div (CALibrated) would become _about_ .25 sec/div if the VARiable knob was rotated fully counterclockwise. Try operating the red VARiable knob and note that it has a click stop in the calibrated position.

CAL

X-Y

ONLY

Many scopes have a X10 (or X5) horizontal magnifier control (explained in Part III). The printed values on the SEC/DIV knob only apply when this control is turned off. Set this control so that the printed values on the SEC/DIV knob will indicate the true sweep time/division for the displayed trace.

2213: Push in the red VARiable knob to turn off the X10 magnifier.

---

*This knob may vary in color and position on different scopes but is usually located near the SEC/DIV knob and marked VAR, VARIABLE or CAL.

(b) If your answer to (a) exceeds 0.2 divisions, the DC balance needs adjustment. Ask your lab supervisor to check it.

| 3.72 | Check the DC balance for channel 2.

Does the CH2 DC balance on your scope need adjustment? _____

(a) do   (b) can

(c) cannot, because the differential mode (using both CH1 and CH2 inputs) is necessary to display v_3

| 4.67 | The phase shift network requires a +12 volts and a −12 volts DC power supply. Make the correct adjustments to your dual power supply* and then connect the power supply to the network. Always turn the power supply output OFF while making or removing connections to the circuit being used.

Using CH1 on your scope, observe the sine wave output of the function generator with f = 1kHz. Ajust the function generator for approximately 3 volts peak-to-peak with no DC offset.

Now connect terminals 1 and 2 of the phase shift network to the oscillator as shown in frame 4.66.

---

*If a dual-supply is not available, use two single supplies connected as follows:

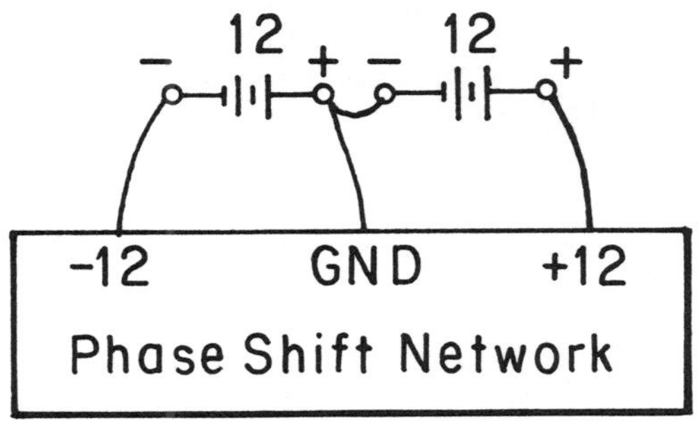

4) 922: CH1   CAL X-Y: CH2

(Omit this frame if your scope does not have a BEAM FINDER.)

1.73 Occasionally, due to the way the horizontal (X) and vertical (Y) POSITION knobs are set or due to the way the vertical (Y) sensitivity is set, the oscilloscope beam will be deflected from the center so much that it goes off the scope screen. When this happens, the BEAM FINDER button, located near the FOCUS knob, will help locate the beam. When the BEAM FINDER button is depressed and held in, the entire range over which the beam can be deflected (on and off the scope screen) is reduced to the size of the scope screen. To locate a beam that has been deflected off screen

(a) Depress and hold in the BEAM FINDER button. If no beam is now visible, turn the INTENSITY control up until the beam appears. Now, as best as possible, center the beam on the scope screen with the appropriate POSITION knobs, then release the BEAM FINDER button.

(b) Now adjust the POSITION knobs to move the beam to its desired position. If necessary, adjust the vertical sensitivity control.

Turn the INTENSITY control on your oscilloscope down until the spot in the center is no longer visible. Turn the horizontal and vertical POSITION knobs counterclockwise as far as possible. Using the BEAM FINDER, locate and recenter the spot.

2.70 Set the TRIGGERING SOURCE switch to EXT.

Set the sweep rate to .5 sec/div (calibrated).

To move across the screen from the leftmost grid line to the rightmost

grid line, the spot should take _____ sec.   (a)

Use a watch to observe the time required for the spot to cross the

screen. Observed time _____ sec.   (b)

(Note that the spot actually goes a little beyond the grid lines when it sweeps across. This is really of no concern since the calibrated sweep rates apply to divisions marked on the screen.)

If the maximum deflection of the horizontal line exceeded 0.2 divisions when you rotated VOLTS/DIV (calibrated) from the maximum to minimum value, the CH2 DC balance needs adjustment. Ask your lab supervisor for help.

Checking the Calibration of the Time Base

| 3.73 | If the scope is to be used to make accurate time measurements, the TIME BASE must be properly calibrated. We will display on the scope screen a 60 Hz sine wave and compare the period as measured using the scope with the expected period of $T = \frac{1}{60}$ sec. If the period measured using the scope differs substantially (by more than 4%) from $\frac{1}{60}$ sec., then the TIME BASE of the scope (is/is not) _____ properly calibrated. (a)

| 4.68 | On the figure below, show the connections to the scope necessary to measure the phase of v_2 relative to v_1 by the dual trace method.

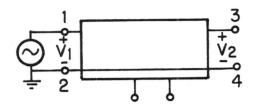 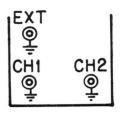

1.74 Investigate the effect of operating the intensity and focus controls. Then adjust these controls to give a small sharp spot. If no spot is present on the screen, the _____ control should be (a) rotated _____. If the spot is fuzzy, the (b) _____ control should be adjusted. (c)

Investigate the effect of operating the position controls.
To move the spot down, the (CH1/CH2) _____ vertical position (d)
control should be rotated _____. (e)
To move the spot to the right, rotate the _____ (f)
_____ control _____. (g)

(a) 5 sec. (.5 sec/div x 10 div)

(b) If your answer is appreciably different from 5 seconds make sure the VARiable (X1 - X10) knob is in the proper position. If your answer is still appreciably off, ask your instructor to check the calibration of your time base.

2.71 If the spot takes 2 seconds to cross the screen, the sweep rate should be set at _____. After writing your answer, verify it experimentally.

3.74 Standard electrical wall sockets provide a 60 Hz AC source, but the voltage from these sockets is much too high ($\approx$ 117 volts). Use a filament transformer* to reduce the voltage to below 15 volts. How many divisions on the scope will be required to display 3 cycles of the waveforms if the SEC/DIV

is set to 5 ms/div (calibrated)? _____ Verify this experi-   (a)
mentally.

The number of divisions you measured on the scope for 3 cycles of the waveform may not agree exactly with the number you calculated above. Calculate the percent error for the TIME BASE. (See frame 3.63 if you do not remember the percent error formula.) If you cannot display a full 3 cycles of the waveform on your scope, calculate the number of divisions required to display 2 cycles and compare this calculation with the measured number of divisions necessary to display 2 cycles.

Percent error of TIME BASE _____   (b)

---

*The 6.3 VAC output on a high-voltage power supply may be used if available.

---

terminal 1 to CH1 input, 3 to CH2 input, 2 or 4 to scope ground.

4.69 In preparation for measuring phase shift of the network, use a 1000 Hz sine wave source to calibrate the time axis for 20°/division with $t = 0$ at the left edge of the graticule. Do this as accurately as you can.

1.75 The vertical (Y) sensitivity is adjustable by means of two knobs. One
knob (usually gray or black) is labeled VOLTS/DIV and changes the sensitivity in steps. The other knob (usually red and in the center of the
VOLTS/DIV knob) is labeled VARiable or CALibrated and allows the sensitivity to be varied continuously between steps. Examine the vertical (Y)
sensitivity control on your scope. Note that the VOLTS/DIV knob has
a dial surrounding it with the vertical sensitivity scales (S_v) printed
on the dial. Note that on the S_v scale, some of the numbers have a small
"m" printed under them. This "m" means that the S_v scale setting is in
MILLIVOLTS/DIV rather than VOLTS/DIV. The S_v scale on the vertical (Y)

sensitivity control ranges from the smallest setting of _____(a)

to the largest setting of _____.   (b)

.2 sec/div

2.72 Set the sweep rate to .5 sec/div.

922
ONLY

Observe the effect of rotating the red X1 - X10 knob on the sweep rate.

Rotating this knob clockwise _____ the speed at which   (a)
the spot moves across the screen.

The minimum time required for the spot to cross the screen (grey knob

still set to .5 sec/div) is _____. This time   (b)

is (faster/slower) _____ by a factor of _____   (c,d)

than the time required for the spot to cross the screen when the red
X1 - X10 knob is in the calibrated position.

CAL
X-Y
ONLY

Observe the effect of rotating the red VARiable knob on the sweep rate.

Rotating this knob counterclockwise _____ the speed at (e)
which the spot moves across the screen.

The maximum time required for the spot to cross the screen (knob still

set to .5 sec/div) is _____.   (f)

(a)   3 cycles x $\frac{1}{60}$ sec. per cycle x $\frac{1 \text{ division}}{5 \text{ msec}}$ = 10 divisions

(b)   If your error is greater than 4%, then the TIME BASE calibration on your scope may need adjustment. If you think your TIME BASE needs calibration adjustment, call your lab supervisor to verify your measurements.

3.75   If the TIME BASE calibration has been properly adjusted, the calibration should be correct for all settings of SEC/DIV. How many cycles of the 60 Hz waveform should be observed for a setting of 10 ms/div?

_____

Verify this.

4.70   Place a check after each of the following items which you did correctly:

(a)   Network connected to source? ☐   (b)   trigger SOURCE set to INTernal ☐

(c)   CAL X-Y only: select trigger source from CH1 ☐

(d) v_1 connected to CH1 input and v_2 connected to CH2 ☐

(e)   vertical mode set to display CH1 only ☐

(f)   fine, sharply focused trace ☐

(g)   peak amplitude 3 or more divisions ☐

(h)   SEC/DIV (both grey and red) set so that trace crosses axis at 9th division ☐

(i)   horiz. position set so t = 0 here ☐

(j)   trigger level set so trace starts here ☐

(k)   if the CH1 vertical input is grounded (with the AC-DC-GND switch), the trace is centered ☐

If you missed any of these points, go back to frame 4.69 and try again.

T922:   (a) 2 millivolts/div   (b) 100 volts/div
OTHERS: smallest and largest settings vary between scopes. Smallest
         settings usually 2-10 mV/div, largest 5-20 volts/div.
         Be sure to include units in your answers.

1.76

922
AND
MOST
TEK-
TRONIX
SCOPES

Examine the dial surrounding the grey knob on the
vertical (Y) sensitivity control. Note that
behind this clear plastic dial is a grey background
with 2 clear "windows" in it. Each of the "windows"
displays a number from the VOLTS/DIV scale imprinted
on the clear dial. In the figure to the right,
what two numbers are being displayed?

_____ These two readings (a)

differ by a factor of _____ (b)

Turn the vertical (Y) sensitivity control to
several different positions. At each position,
record the two VOLTS/DIV values that appear in the
windows. Do the two readings always differ by the

same factor? _____ (c)

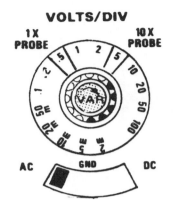

VOLTS/DIV

1 X PROBE   10 X PROBE

AC   GND   DC

(a) increases   (b) approximately .5 sec (with the red knob fully
                           clockwise)

(c) faster   (d) 10

(e) decreases   (f) approximately 15 sec (with the red knob fully
                           counterclockwise)

2.73 Again with the sweep rate calibrated, observe the effect of changing the
      sweep rate. Observe that

      (a) at slow sweep rates the spot can be seen moving across the screen

      (b) as SEC/DIV is decreased the spot gradually becomes a line, but
           the line flickers on and off

      (c) as SEC/DIV is further decreased, the trace becomes a steady line
           without any noticeable flicker

What is the setting of the SEC/DIV switch at the boundary between condi-

tion (b) and condition (c)? _____

3.76 Using the 60 Hz waveform, adjust the SEC/DIV knob and associated variable knob so that the sweep rate is 1.67 ms/div.

setting of SEC/DIV _____

number of cycles on screen _____

4.71 Now measure the phase angle of v_2 with respect to v_1 at 1000 Hz.

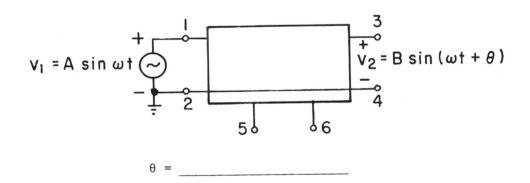

$v_1 = A \sin \omega t$

$v_2 = B \sin (\omega t + \theta)$

$\theta = $ _____

(a) .5 volts/div and 5 volts/div   (b) 10

(c) Yes, the readings always differ by a factor of 10

| 1.77 | When the scope is used with a coaxial cable lead (or a X1 probe), the correct VOLTS/DIV value for the vertical sensitivity (S_v) is the <u>smaller</u> of the two values in the windows. Note on your scope that the leftmost window will always have the smaller of the two values. Note also that this window is labeled "X1 PROBE". The larger value in the other window is used with a X10 probe, as will be discussed in Laboratory Part III. Note on your scope that this window is labeled "X10 PROBE". For the figure above if a coaxial cable lead is used, what is the correct VOLTS/DIV value? _____ |

922
AND
MOST
TEK-
TRONIX
SCOPES

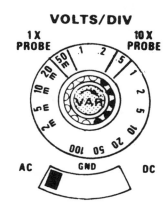

VOLTS/DIV

1X PROBE   10X PROBE

VAR

AC   GND   DC

OTHER
SCOPES{ When the scope is used with coaxial cable (or X1 probes), the correct VOLTS/DIV values for vertical (Y) and horizontal (X) sensitivity will be that indicated by the VOLTS/DIV knobs. (A X10 probe, introduced in Part 3, will require scaling the indicated VOLTS/DIV values.)

---

roughly 2 msec/div

Triggering Controls

| 2.74 | Examine the triggering LEVEL control on your scope. Note that there are no calibration markings on the knob. Turn the LEVEL control counterclockwise as far as possible. Now turn the LEVEL control clockwise as far as possible. Now try to center the knob between these two positions. When the knob is centered, the LEVEL is set to approximately 0. Turning the knob clockwise increases the LEVEL setting and turning the knob counterclockwise decreases the setting. |

If the LEVEL control is initially centered, which direction should it be turned to cause the scope to trigger at a positive point on the triggering

waveform? _____ At a negative point on the   (a)

triggering waveform? _____   (b)

922 ONLY: You should have set the grey knob to 2 <u>msec</u> and adjusted the red knob to obtain exactly 1 cycle on the screen.

CAL X-Y ONLY: You should have set the SEC/DIV knob to 1 <u>msec</u> and adjusted the red variable knob to obtain exactly 1 cycle on the screen.

3.77 Reset your scope controls for calibrated time base operation. Connect a sine wave oscillator to the CH1 input and set it to 1500 Hz.

If the oscillator output frequency was exactly the same as the dial setting, the period of the sine wave would be _____. (a)

Now measure the actual period of the sine wave as accurately as you can. Choose an appropriate setting for SEC/DIV and adjust the trigger controls to facilitate this measurement.

    Measured period _____   (b)

    SEC/DIV setting _____   (c)

$\theta = -129^{\circ}$ (any answer in the range -126° to -132° is acceptable)

If your answer checks within the prescribed tolerance, skip the rest of this frame.

4.72 If your answer is wrong, check the following:
- (a) The triggering controls should be set exactly as they were in frame 4.69.
- (b) The network should be connected as in frame 4.68.
- (c) Both of the vertical input switches should be in the DC position (or both in the AC position).
- (d) With the CH2 input grounded (using the AC-DC-GND switch) the CH2 trace should be centered.
- (e) v_2 is negative at $t = 0$, so the sign of θ is _____.
- (f) The magnitude of θ is _____ div x _____ degrees/div.

If you still can't get -129° ($\pm 3^{\circ}$), your oscillator may need calibration. Ask the supervisor for help.

50 mV/div (the smaller of the two readings)

1.78 Turn the vertical (Y) sensitivity control on your scope to its most sensitive position. Assuming coaxial cable leads, what vertical input voltage would be required to produce a full scale deflection of 4 cm? _____ Now turn the vertical (Y) sensitiv- (a) ity control to its least sensitive position. What vertical input voltage would be required to produce a full scale deflection? _____ (b)

(922: Be careful to read S_v from the correct window.)

(a) clockwise   (b) counterclockwise

2.75 We will now use the external trigger input on the scope. On some scopes, two ranges of external trigger inputs may be selected. On one range, the external trigger input is fed directly to the time base and on the other range, the signal may be divided by 10 before it is fed into the time base (see Fig. 5-3 on page 234). On the 922 scope, these two ranges are selected by setting the TRIGGERING SOURCE switch to EXT or EXT/10. On the Tektronix 2213 these two ranges are selected by the DC and DC/10 settings on the TRIGGER COUPLING switch. If your scope has two ranges for the external trigger input, determine how these ranges are selected and select the range which does not divide the input by 10.

922 ONLY

When the TRIGGERING SOURCE switch is set to EXT, the LEVEL control may be adjusted to any value between −.5 volts and .5 volts. Turning the LEVEL control counterclockwise as far as possible sets LEVEL to _____ (a) and turning the LEVEL control clockwise as far as possible sets LEVEL to _____. When the SOURCE switch is set to $\frac{EXT}{10}$, the LEVEL (b) control may be adjusted to any value between −5 volts and 5 volts. When the triggering point on the external signal is larger in magnitude than .5 volts but smaller than 5 volts, the TRIGGERING SOURCE switch must be set to _____. If the triggering point is smaller in mag- (c) nitude than .5 volts, TRIGGERING SOURCE should be set to _____ (d) for best results.

(a) 1/1500 = 0.667 ms

(b) Any answer in the range of 0.64 to 0.71 is acceptable (this allows for possible errors in the oscillator dial calibration and for errors in the scope sweep rate).

(c) 0.1 ms/div for 1-1/2 cycles on the screen. (Or you could have used 0.2 ms/div and measured the time for 3 periods of the sine wave.)

Checking Horizontal Amplifier Sensitivity

3.78 In this step you will measure the horizontal amplifier sensitivity. Select the X-Y mode on your scope.

922 ONLY: Set the X1-X10 knob for X1 operation.
CAL X-Y ONLY: Set the horizontal sensitivity to 1 volt/div (calibrated).

Center the spot. Apply a 1 kHz sine wave to the X input and adjust the amplitude until the horizontal line is exactly 10 divisions. Use a digital voltmeter to measure the amplitude of the input voltage.

Give an equation which expresses the horizontal sensitivity in terms of the measured RMS input voltage _____.   (a)

Using this equation, the measured horizontal sensitivity is _____.(b)

Record the value of θ which you measured at 1000 Hz. in Table 4-1 (p. 226).

4.73 When measuring phase shift by the dual trace method, the time axis of the scope must be recalibrated each time the frequency is changed because _____

_____

160

(a)   Required Voltage = smallest volts/div (from 1.72) x 4 div

(b)   Required Voltage = largest volts/div x 4 div

For T922:   (a)   8 millivolts   (b)   40 volts

1.79   The voltage gain is calibrated only when the red* VARiable knob is in the CALIBRATED position. To be in the calibrated position, the VARiable knob must be turned as far as possible in the clockwise position (↻). A definite click will be felt when going from calibrated to uncalibrated and back. (Try it and see.) The numerical values of VOLTS/DIV printed around the knob apply only when the VARiable knob is in the CALIBRATED position. If the scope is to be used for making accurate voltage measurements, the red VARiable knob that is associated with the VOLTS/DIV

knob should be rotated fully _____.

---

*On some scopes this knob is grey or black

(a)   -.5 volts   (b)   .5 volts   (c) $\frac{EXT}{10}$ (d)   EXT

2.76   When NORMAL triggering mode is used, a signal is required to trigger the

sweep. This signal may come from the _____, the _____

or the _____.   (a)

Select the NORMAL triggering mode (turn off AUTO).

922 ONLY { We want the sweep to be triggered by an external input in the range -5 to +5 volts, so the trigger SOURCE switch should be set to _____. (b) Set the SOURCE switch to this position.

CAL X-Y ONLY { We want the sweep to be triggered by an external input, so the trigger SOURCE switch should be set to _____. Set the SOURCE switch (c) to this position. Set the trigger COUPLING to DC.

Set the SLOPE to trigger on the positive (+) slope of the waveform, the LEVEL a little to the right of the center, and SEC/DIV to .2 sec (calibrated).

Does the sweep trigger when there is no external trigger input? _____ (d)

Momentarily apply a small positive DC voltage (about 10 volts) to the external trigger input (922 Only: the same X connector used for the horizontal amplifier input in Lab Part 1). Describe what happens.

_____ (e)

If you have trouble getting the sweep to trigger, try adjusting the LEVEL.

(a) $2\sqrt{2}$ x RMS input voltage/10 divisions

(b) 922: Your answer should be between 750 and 800 mv/div.
CAL X-Y: 1.0 volts/div. If error exceeds 4% your scope may need
calibration. Ask the supervisor to check.

| 3.79 | CAL X-Y ONLY: With the same VOLTS/DIV setting, select the X10 position of the horizontal magnifier.

922 ONLY: Set the X1-X10 knob for X10 operation.

ALL: Check the horizontal sensitivity in the X10 position.

Measured horizontal sensitivity _____

Changing frequency changes the number of degrees per division

---

| 4.74 | Check the vertical centering of the traces with the input switches grounded to make sure that the trace hasn't drifted.

Now measure the phase shift at 100 Hz. θ = _____

clockwise (past the click stop, so that it is in the calibrated position)

1.80 Examine the vertical (Y) input connector on the scope.

The ungrounded or "high" side of the input should be connected to the
(center, outer) _____ conductor.   (a)

For this part, use a coaxial cable with clips on one end.

Connect this cable to the vertical (Y) input. Which of the coaxial cable
conductors is now grounded? (center, outer) _____.   (b)

The ungrounded conductor is terminated in a clip with a (red, black)
_____ insulator. (Use the continuity checker if you   (c)
can't tell by inspection.)

(a) vertical amplifiers (internal trigger), AC line, external trigger
     input

(b) $\frac{EXT}{10}$ (c) EXT

(d) NO   (e) The spot crosses the screen once (since the sweep
                    is triggered once).

2.77 Without changing the SEC/DIV, suppose you were to connect a sine
wave with a frequency of 0.2 Hz to the external trigger input. Sketch the
sweep waveform (approximately) which should be generated by the TIME
BASE. (This waveform does not appear on the scope face, of course.)

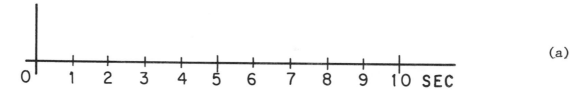

(a)

The sweep should be triggered once every _____ seconds and the   (b)
spot should take _____ seconds to cross the screen. Verify this   (c)
by observation. (If the sweep fails to trigger, turn up the amplitude
on the function generator.)

163

922:   95 to 105 mv/div.

CAL X-Y: approximately 0.1 volts/div.   ($\approx$.2 volts/div for 2213)

3.80   Adjust the horizontal sensitivity to 400 mv/div. What peak-to-peak

horizontal deflection did you use? _____ (a)

What was the voltmeter reading for this deflection? _____ (b)

$-14°$ (an answer in the range $-11°$ to $-17°$ is acceptable)

If you didn't get the correct answer, try again.

When you get a value in the correct range, record it in Table 4-1 (p. 226).

4.75   Now measure the phase shift at 300 Hz, 3000 Hz and at 10,000 Hz.

θ at 300 Hz = _____ (a)

θ at 3000 Hz = _____ (b)

θ at 10,000 Hz = _____ (c)

(a) center   (b) outer   (c) red

Deflection of the Spot by DC Inputs

1.81 We are now ready to study the effects of applying DC signals to the
scope input terminals.

Ask your instructor for information which explains the functions
of the controls on your DC power supply. If you have a dual power
supply continue with this frame, otherwise, turn to frame 1.82.

The equivalent circuit
for the dual power supply
output is:

$-(0-20V)$ V_2 COMMON V_1 $+(0-20V)$

CASE GROUND

Note that the ground terminal is a case ground and is not connected to
any of the three output terminals.

If the magnitude of V_1 is adjusted to 4 volts and the magnitude of V_2
is adjusted to 10 volts, what voltage would you expect to measure

at the $+(0-20V)$ terminal relative to the GROUND terminal?_____ (a)
at the $-(0-20V)$ terminal relative to the GROUND terminal?_____ (b)
at the $+(0-20V)$ terminal relative to the COMMON terminal?_____ (c)
at the $+(0-20V)$ terminal relative to the $-(0-20V)$ terminal?_____ (d)
at the $-(0-20V)$ terminal relative to the COMMON terminal?_____ (e)

Check your answers above frame 1.83 (two frames ahead).

(a)

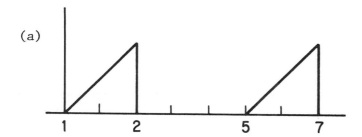

(b) 5 sec (once every cycle)

(c) 2 sec (.2 sec/div x 10
     div)

2.78 Make the necessary changes so that the sweep will trigger 10 times a
second and the spot will take 50 ms to cross the screen. Record the dial
settings that you had to change. _____

_____

(a)   10 divisions

(b)   1.41 volts (RMS)

THIS IS A GOOD PLACE TO TAKE A BREAK

Use of the Probe

3.81 To reduce loading of the circuit, the probe should be used for making measurements on high-impedance circuits and for observation of high-frequency
signals.

Recall the equivalent circuit of the probe and scope input:

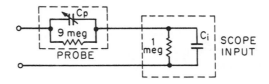

In order to compensate the probe so that the probe attenuation is independent of frequency and the probe will not distort waveforms, we must

_____.

(a)   −43 ± 3° at 300 Hz

(b)   +112 ± 3° at 3 kHz   (c)   +36 ± 3° at 10 kHz

If correct, record your answers in the table.

If not correct, try again.

Is the trace centered when both vertical inputs are grounded?

THIS IS A GOOD PLACE TO TAKE A BREAK

TURN TO FRAME 4.76

1.82 | The equivalent circuit for the single power supply output is:

$+ \quad V \quad -$

CASE GROUND

Note that the ground terminal is a case ground and is not connected to either the + or - output terminal.

If the amplitude control on the power supply is adjusted so that the magnitude of V is 4 volts, what voltage would you expect to measure

    between the + terminal and ground?_____ (a)

    between the - terminal and ground?_____ (b)

    at the - terminal relative to the + terminal?_____ (c)

---

generator frequency changed to 10 Hz and TIME/DIV (SEC/DIV) changed to 5 ms.

2.79 | We have been triggering the sweep from a high amplitude external source. Now set the TRIGGERING SOURCE so that the sweep will trigger from the vertical input.

922 ONLY { When the CH1 vertical mode button is depressed, the sweep will trigger off the CH1 input and when the CH2 button is depressed, the sweep will trigger off the CH2 input.

CAL X-Y ONLY { Most scopes have a triggering control with CH1-CH2 settings which allows either the CH1 or CH2 input signal to be used as a trigger source independent of which channel is selected by the vertical MODE switch for display on the screen. If the input to CH1 is to be displayed and used as the trigger input the vertical MODE switch and the trigger internal source switch must both be set to CH1.

Until further notice, we will use only the CH1 input. Set the appropriate button(s) or switch(es).

Does the sweep trigger when there is no vertical signal present?_____ (a)

If we want a sweep when there is no signal present, the triggering mode

should be set to _____. (b)

If all inputs are disconnected from the scope, the triggering mode is set to NORM, the LEVEL control is set to 0 (centered), how should the SOURCE

switch be set if we want the sweep to trigger?_____ (c)

Verify (b) and (c) by observation.

adjust C_p (so that $C_p = C_i/9$)

3.82 Each time a probe is transferred from one scope to another, the compensation of the probe should be checked. If a 1 kHz (1000 Hz) square wave was observed using the probe and one of the waveforms of Fig. 5-8 (p. 239) was observed, in which case(s) must the compensation of the probe be adjusted?

(a)

Some probes (such as the B & K PR-37) have a switch marked X1-REF-X10. When the switch is in the X10 position, the probe circuit contains the 9 megohm resistor and compensating capacitor described in the preparation. In the REF position, the scope vertical input is grounded and the probe tip is connected to ground through the 9 megohm resistor. In the X1 position, the probe tip is connected directly to the scope vertical input and the probe will work the same as the coaxial cables used earlier.

If your probe has a X1-REF-X10 switch, select the position which will place the 9 megohm resistor and compensating capacitor in the circuit. (b)

Phase Measurement by the Triggered Sweep Method

4.76 (omit this part if your scope does not have a differential input option)

Now we will measure the phase angle of v_3 relative to v_1. For the triggered sweep method, show the connections to the scope required to calibrate the time axis (step 1), and to measure the phase angle (step 2).

STEP 1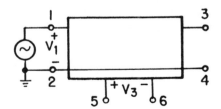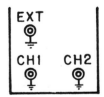

STEP 2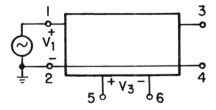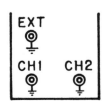

Triggering SOURCE for both step 1 and step 2 should be _____.

If your answer to (a) is +4, continue with this frame; otherwise, turn to the next frame.

1.83 On the equivalent output circuit diagram in frame 1.81 (if you have a dual power supply) or frame 1.82 (if you have a single power supply), draw a resistor between the + output terminal and the ground terminal.

How much current would flow in this resistor? _____ (the answer is <u>not</u> V/R since current flow requires a closed path). If we replace the resistor with a voltmeter with internal resistance R, how much current will flow? _____ If no current is flowing in the meter, it will read _____. Now turn back to frame 1.81 (dual) or 1.82 (single) and try again.

(a)   no   (b)   AUTO   (c)   LINE

THIS IS A GOOD PLACE TO TAKE A BREAK

Displaying Periodic Waveforms

2.80 Set the vertical sensitivity to 10 volts/div and set the scope controls so that the sweep will trigger only when a CH1 vertical input is present.

If a 0.2 Hz sine wave were connected to the vertical input, the sweep would be triggered once every _____ sec.   (a)

The (CH1) AC-DC-GND switch should be set to _____

because _____.   (b)

CAL*
X-Y   { Similarly, the trigger coupling should be set to _____.   (c)
ONLY

Make the necessary adjustments of the trigger controls and verify your answers by observation. (Don't try to get a sine wave on the screen; just observe the triggering of the sweep.)

_____

*Omit for 2213.

169

(a) The compensation of the probe must be adjusted in cases (a) and (c) because the square wave is distorted.

(b) You should have selected the X10 position.

3.83 For the remainder of this course, we will use the term probe to mean a 10X probe, or a probe with the switch set to the 10X position. Connect your probe to the CH1 input.

On your scope, locate the jack marked PROBE ADJ or CAL (just to the right of the POWER switch on the 922 scope). This jack provides a square wave of approximately 1000 Hz with a peak-to-peak amplitude of about 500 mV (1 volt on some scopes). In preparation for calibrating the probe, display this calibrator signal by touching the PROBE ADJ jack with the metal pin at the end of the probe.* Adjust the scope controls to display about 5 cycles of the waveform with 5 divisions peak-to-peak deflection.

If you have a Tektronix P6006 or similar probe, turn to frame 3.85.

If you have a Tektronix P6108 or similar probe, turn to frame 3.86.

If you have a B&K PR-37 or similar probe, continue with the next frame.

---

*On high frequency scopes, it may be necessary to also connect the probe ground lead to the scope ground.

Triggering SOURCE to EXTernal

(step 1) v_1+ to CH1 input and EXTernal trigger input

(step 2) v_3+ to +INPUT (CH1) and v_3- to −INPUT (CH2); v_1+ to EXTernal trigger input

function generator ground to scope in both cases

4.77 Use a 2000 Hz sine wave (amplitude<5 vp-p) to calibrate the time axis for 20°/division.

Now measure the phase angle of v_3 with respect to v_1.

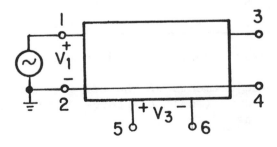

$$\theta = \underline{\qquad\qquad}$$

Answers to 1.81 (dual supply) (a) 0 volts   (b) 0 volts (since none of the output terminals are connected to ground)   (c) +4 volts (d) +14 volts   (e) -10 volts.

Answers to 1.82 (single supply) (a) 0 volts   (b) 0 volts (since neither the + or - terminal is connected to ground).   (c) -4 volts.

1.84   [CAL X-Y ONLY: Ground the horizontal (X) input using the AC-DC-GND switch.]

Set the vertical (Y) sensitivity to 1 VOLT/DIV (calibrated).

Recenter the spot if necessary.

Connect a small positive DC voltage from the power supply to the vertical (Y) input. (If your power supply does not have a built in meter, connect a DC voltmeter to measure the output voltage.)

Set the vertical AC-DC-GND switch to the proper position and observe the deflection of the spot as the input voltage is varied.

(a)   once every 5 seconds (with SEC/DIV set to 0.2 sec or less)

(b,c)   DC because the input is very low frequency

2.81   We now wish to display a 1000 Hz sine wave on the scope.

If we want to see exactly one cycle of the wave on the scope, we should set the SEC/DIV to _____.   (a)

If we set the SEC/DIV to .2 msec, what would you expect to see on the screen? _____   (b)

Verify your answers experimentally. It may be necessary to adjust the oscillator frequency slightly to get exactly one cycle on the screen for (a).

3.84 | Using a small screwdriver, turn the probe adjusting screw (located above the X1-REF-X10 switch on the B&K probe) no more than one turn in either direction. As the probe is adjusted, observe how the displayed waveform changes from Fig. 5-8(a) to (b) to (c). What circuit element in the probe is being changed? _____. Adjust the probe for correct waveform.

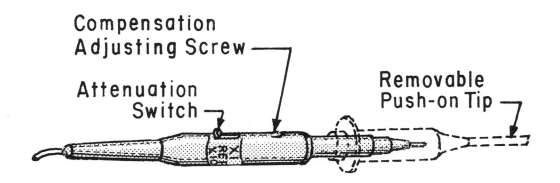

Compensation Adjusting Screw

Attenuation Switch

Removable Push-on Tip

FIG. 3-84. B&K PR-37 PROBE

$\theta = -64^{\circ} \pm 5^{\circ}$

If your answer is within tolerance, skip the rest of this frame.

4.78 | If your answer is wrong, did you

Step 1 (a) connect the network as in frame 4.76. Step 1? _____
       (b) set the vertical mode to CH1? _____
       (c) set the CH1 AC-DC-GND switch to ground and center the trace? _____
       (d) adjust the SEC/DIV (red and grey) knobs and horizontal POSITION knob to obtain the trace shown on p. 4.14? _____

Step 2 (a) connect the network as in frame 4.76. Step 2? _____
       (b) set the vertical mode to differential amplifier operation? _____
       (c) remember both VOLTS/DIV must be the same? _____
       (d) ground both inputs and center the trace? _____

For both steps the triggering SOURCE should have been EXTernal and v_1 should have been connected to the external trigger input. Now turn back to frame 4.77 and try again.

1.85 In this frame and many of the following frames, you will be asked
to predict a pattern or an input voltage <u>before</u> you check your
answer experimentally. Answer the questions and write down your
answers first. Wait to verify your answers experimentally until you
are told to do so.

An input voltage of +3 volts should deflect the spot upward

_____ division. (a)

With the same input voltage, if the input switch is changed to AC,

the deflection should be _____. (b)

After writing your answers, verify them experimentally.

Observed deflection, switch set to DC _____. (c)

Observed deflection, switch set to AC _____. (d)

_____

(a) 0.1 msec/div (since period of sine wave is 1 msec)

(b) 2 cycles of the sine wave (since the spot takes twice as long to
     sweep across the screen)

_____

2.82 (a) With the trace centered on the screen, adjust the TRIGGER LEVEL con-
           trol so that the sweep triggers <u>exactly</u> when the sine wave goes
           through 0. Do this by observing the trace as you adjust the LEVEL;
           the 0 mark on the LEVEL control is not exact.

      (b) Without changing the LEVEL control, investigate the effect of changing
           the TRIGGER SLOPE switch from + to -.

      (c) With the controls still set for 2 cycles of a 1000 Hz sine wave,
           sketch the waveform which you observe with the <u>slope</u> set to -
           (negative).

      (d) Label the time scale.

t (ms)

(Skip frames 3.85 and 3.86. Check your answer above frame 3.87).

3.85 | Loosen the flanged locking sleeve on the probe several turns. Rotate the probe body and tip assembly while holding the base of the probe. CAUTION: Do not rotate more than two turns in either direction. As the probe is adjusted, observe how the displayed waveform gradually changes from Fig. 5-8 (a) to (b) to (c). What circuit element in the probe is being changed? _____

Adjust the probe for the correct waveform and then carefully tighten the locking sleeve.

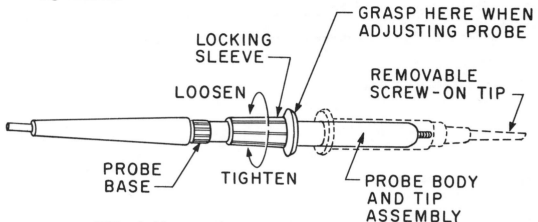

FIG. 3-85. TEKTRONIX P6006 PROBE
Used by permission of Tektronix, Inc.

4.79 | When switching back and forth between CH1 and differential mode operation, the zero line may shift up or down depending on the settings of the vertical position controls. Ground the CH1 and CH2 inputs and observe whether this shift occurs with your scope. To avoid errors in phase shift measurement, what should be done each time you switch from CH1 to differential mode or vice versa? _____

_____ (a)

Now measure the phase angle of v_3 relative to v_1 at 10 kHz.

θ = _____ (b)

(a) 3   (b) 0

If your experimental answer doesn't check, ask your instructor to check
the calibration of your scope.

1.86   Full scale vertical deflection (from center to the top grid line)

is _____ div.   (a)

Investigate the effect of the red variable gain knob (concentric to
VOLTS/DIV) on the deflection. When the red knob is fully counter-

clockwise, _____ volts are required for a full-scale   (b)

deflection so the sensitivity is _____ volts/div.   (c)

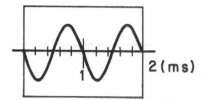

(If you didn't get the picture, go
back and try again. Make sure the
frequency is 1000 Hz. On some scopes,
slight readjustment of the TRIGGER
LEVEL may be necessary so that the
sweep still triggers exactly when the
sine wave goes through zero.)

2.83   If you changed the signal generator to give a square wave of the same amplitude and frequency, sketch what you would expect to see on the scope.
(TRIGGER SLOPE still –)

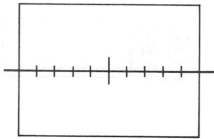

Make the change and check your prediction. Also investigate the effect of
changing the TRIGGER SLOPE when observing the square wave.

(Skip frame 3.86 and check your answer above 3.87).

3.86 Using a small screwdriver, turn the probe adjusting screw (located on the compensation box attached to the BNC connector) no more than one turn in either direction. As the probe is adjusted, observe how the displayed waveform changes from Fig. 5-8(a) to (b) to (c).

What circuit element in the probe is being changed? _____.

Adjust the probe for correct waveform.

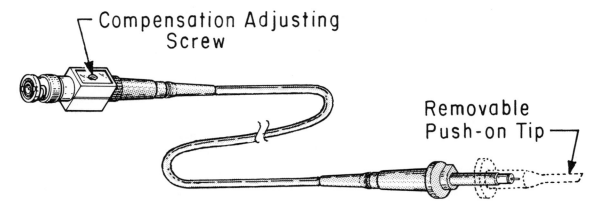

FIG. 3-86. TEKTRONIX 6108 PROBE
Used by permission of Tektronix, Inc.

(a) center the trace with the inputs grounded

(b) $-153^{\circ} \pm 3^{\circ}$

Phase Measurement by the Ellipse Method

4.80 When using the ellipse method, instead of displaying v_1 vs time or v_2 vs time on the screen, we display v_2 vs v_1. Indicate the proper connections for measuring the phase angle between v_2 and v_1 by the ellipse method.

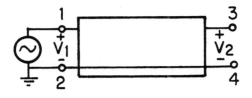   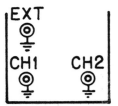

(a) 4 div.   (b) about 10 to 20 volts

(c) to get sensitivity, divide your answer to (b) by 4.

1.87 Change the vertical gain to 5 volts/div (calibrated). Check the cen-
tering of the spot with the <u>input grounded</u> whenever you change the
gain setting. A vertical deflection of 2 divisions should require
an input of _____. Verify your answer experimental-   (a)
ly. Observed voltage _____.   (b)

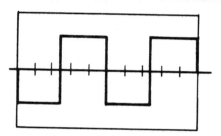

2.84 Change the input to a 800 Hz sine wave and observe the effect of
varying both the grey SEC/DIV control and the associated red
VARiable control (X1 - X10 on 922) over a wide range of sweep rates. Now
set the scope to obtain the following picture:

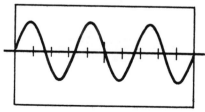

Observed setting of SEC/DIV _____   (a)

Calibrated or uncalibrated? _____   (b)

Rotating the red VARiable control counterclockwise (increases/
decreases) _____ the time/div.   (c)

The probe capacitance is being changed. Recheck the compensation of the probe. Make sure that the corners of the square wave are square with no rounding or overshoot.

3.87 The probe attenuates the input signal by a factor of _____. (a)
Still using the probe to observe the 500 mV calibrator output and with the vertical deflection still 5 div peak-to-peak, the observed setting

of the VOLTS/DIV switch (left window on 922) is _____. (b)

When the probe is used, to get the true value of volts/div, the setting of the VOLTS/DIV switch (left window on 922) must be (multiplied/divided)

_____ by _____. (c,d)

922
AND
MOST
TEK-
TRONIX
SCOPES

The right window displays a volts/div value _____ times (e)
greater than the left window. Therefore, when using the probe, the true VOLTS/DIV value can be obtained from reading the number in the

_____ window. (f)

---

v_1+ to external horizontal (X) input v_1- or v_2- to scope ground

v_2+ to vertical (Y) input

4.81
1. Set up the scope to measure the phase shift of the network by the el-lipse method. Use the same phase shift network and oscillator that you used for the dual trace method. Do NOT use the Webb mask until you are told to do so.

2. Set the input switch(es) to DC. Adjust the oscillator and scope con-trols to obtain a ellipse which occupies almost the entire height and width of the screen.

3. Vary the oscillator frequency from 20 Hz to 40 kHz and observe the effect on the ellipse.

    As the frequency approaches zero, the phase shift of the network
    approaches _____. (a)

    At what frequency is the phase shift 180°? _____ (b)

(a) 10 volts   (b) 10 volts

1.88 Change vertical gain to 0.5 volts/div.

An input of 2 volts should now deflect the spot _____ (a)
divisions.

Verify your answer experimentally. Observed deflection

_____ (b)

Now turn the red VARiable volts/div knob back and forth. As the
red knob is turned counterclockwise, the deflection of the spot

_____ so the effective VOLTS/DIV (c)

_____. (d)

(a) 922: .5ms CAL X-Y: .2ms   (b) uncalibrated   (c) increases

---

2.85 Display several cycles of a 1000 Hz square wave on the screen. We
now want to adjust the amplitude of the function generator output to
obtain a 12 volt peak-to-peak (6 volt peak) square wave.

What setting of VOLTS/DIV (vertical sensitivity) should be used?

_____ (a)

What should the peak-to-peak deflection be? _____ (b)

Now adjust the generator amplitude control to obtain a 12 volt
peak-to-peak output. (The amplitude control is not marked in volts,
so adjust it while watching the scope.)

(a) 10   (b) 10 mv/div.   (c) multiplied

(d) 10   (e) 10   (f) right

3.88   Next we will observe the loading effect of the scope without and
then with the probe.

Connect the circuit shown below. Set the oscillator frequency to
100 Hz. Observe the oscillator output with the scope (use the
coaxial cable lead, not the probe). Adjust the oscillator to obtain
an output of V_1 = 10 volts peak-to-peak as accurately as you can.
(Leave the network connected to the oscillator when you measure the
oscillator output since the output may change when you disconnect
the network.) What VOLTS/DIV setting did you use?

_____   (a)

How many divisions peak-to-peak deflection did you obtain?

_____ .   (b)

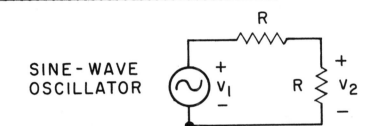

Use R ≈ 500 kilohms.
Obtain a pair of precision
resistors from the super-
visor.

(a) 0   (b) 1590 Hz ± 50 Hz

If both of your answers are correct, turn to frame 4.83

Otherwise, continue with this frame.

4.82   Check your circuit as follows:

Oscillator ground connected to scope ground? ☐

Other oscillator terminal (v_1+) connected to the external horizontal (x)
input? ☐

Output terminals from phase shift network (v_2) connected to the
vertical (Y) input? ☐

Input switch(es) set to DC? ☐

Oscillator output amplitude or horizontal sensitivity set to
give desired trace width on screen? ☐

Now turn back to frame 4.81 and try again.

(a) 4 div   (b) 4 div   (Did you remember to center the spot?)

(c) decreases   (d) increases

1.89   If the gain is set to 2 volts/div, then how many volts would be required

to deflect the spot <u>downward</u> 3 divisions? _____ (a)

Verify your answer experimentally. (Make any necessary changes to the

connections on the power supply.) Observed voltage _____ (b)

(a) 2 volt/div (calibrated)   (b) 6 div.

2.86   Adjust the generator output to obtain a 3.5 volt peak-to-peak sine wave:

(a) First do this using 2 VOLTS/DIV as above.

(b) Change VOLTS/DIV to get the largest vertical deflection possible
and still have the peaks of the sine wave within the graticule.
(The red knob must remain in the CALIBRATED position of course.)
Now readjust the generator output if necessary to obtain 3.5 volts
peak-to-peak.

VOLTS/DIV setting _____ peak-to-peak deflection

_____.

(c) In which case, (a) or (b), was accurate adjustment of the sine
wave amplitude easier to make? _____

(d) In general, if we are trying to adjust the vertical input voltage
as accurately as possible, we should

_____

_____

(a) 2 VOLTS/DIV   (b) 5 div.

You also could have used 5 volts/div and 2 div or 10 volts/div and 1 div, but this would not be as accurate. If you didn't use 2 volts/div and 5 div, check your result using these values (adjust the position control as required).

3.89 Still using the coaxial cable lead, measure the peak-to-peak value of V_2. Explain why V_2 differs from the theoretical value of 5.

_____ (a)

Now measure V_2 for f = 1 kHz, 10 kHz and 40 kHz. <u>Each time you change frequency observe the oscillator output (V_1) to make sure that it is still 10 volts peak-to-peak</u> (with the network connected).

Complete the following table:

f(Hz)	V_1	V_2
100	10	
1K		
10K		
40K		

← peak-to-peak voltages

(b)

Explain why V_2 decreases with increasing frequency. _____

_____ (c)

(If you can't explain it, go back and review frame 3.7).

Record your answer to 4.81 in Table 4-1 (p. 226).

4.83 Before continuing with the phase measurements, we will examine a possible source of error in our measurements. In addition to amplifying the signal, the scope amplifiers may also introduce phase shift.

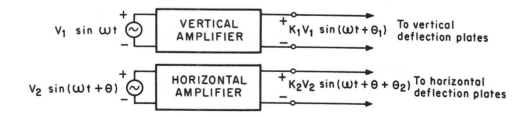

In the above diagram, the phase shift introduced by the vertical amplifier is _____ and that <u>introduced</u> by horizontal   (a)
amplifier is _____   (b)

What relation must hold between these angles if we want to introduce no error in the phase shift measurement by the ellipse method?

_____   (c)

182

(a,b) -6 volts (Note that experimental verification requires reversing the connections to the DC supply so that the + terminal is grounded. A dual power supply may use the -(0-20V) output terminal and the COMMON terminal will remain at scope ground.)

1.90   If it is desired that an input voltage of 15 volts deflect the spot two divisions upward, the VOLTS/DIV knob should be set to _____ (a) and the red VARiable knob should be set to the (calibrated, uncalibrated) _____ position.   (b)

Verify your answers experimentally.
   Observed setting of VOLTS/DIV _____.   (c)
   Observed setting of red VARiable knob _____.   (d)

(b) .5 volts/div, 7 div. (If you used 3.5 div, go back and try again.)
(c) accurate adjustment is easier in (b)
(d) use the largest possible vertical deflection

2.87   Compute the SEC/DIV switch setting required to display 2 cycles of a 400 Hz triangular wave. _____   (a)
Verify your answer by observation.

Set the triggering mode to NORMal.

With 2 cycles of a 400 Hz triangular wave displayed on the screen, investigate the effect of adjusting the LEVEL control with the SLOPE set to + and also with the SLOPE set to -.

Why does the trace disappear if the LEVEL control is turned too far to either side of zero? _____   (b)

_____

(a)   The input impedance of the scope loads down the circuit.

(b) V_2 = 4, 4, 2.4, 0.8 for a scope with 30pf input capacitance and a 3
      ft. coaxial cable. (If your answers don't check within ± .2 volts,
      try again. Each time you change the oscillator frequency, make sure
      that you readjust the oscillator output for 10 volts peak-to-peak.)

(c)   The loading increases with increasing frequency due to the shunt
      capacitance.

3.90   Repeat the measurements made in frame 3.89 using the probe instead of
       the coaxial cable lead. What change should be made in the setting of
       VOLTS/DIV to compensate for the probe attenuation? _____ (a)

f(Hz)	V_1	V_2
100	10	
1K	10	
10K	10	
40K	10	

← peak-to-peak
  voltages

(b)

Compare with the results of 3.89 and explain the difference.

_____ (c)

_____

(a) θ_1 (b) θ_2 (c) $\theta_1 = \theta_2$

4.84   How could you test to see if the phase shift produced by the horizontal
       and vertical amplifiers is the same?

_____

_____

_____

_____

If you want a hint, turn to frame 4.85.

If you know the answer, turn to frame 4.86.

(a,c)   5 volts/div   (b,d)   uncalibrated

1.91

922
ONLY

We are now going to investigate horizontal deflection of the spot. The horizontal sensitivity is adjustable by a red knob labeled X1 ⬌ X10 located in the TIME BASE section of the scope. Make sure that this knob is set in the X1 position (rotated fully counterclockwise).
Apply a DC voltage to the horizontal input and observe the deflection of the spot as the voltage is varied. Apply enough voltage for full-scale deflection of the spot and determine the horizontal sensitivity.

Measured voltage? _____   (a)

S_h = _____ volts/div.   (b)

CAL
X-Y
ONLY

The horizontal sensitivity is adjustable by the CH1* VOLTS/DIV knob. Make sure the corresponding red VARiable control is in the calibrated position.
If the gain is set to 2 volts/div. then how many volts would be required to deflect the spot 4 divisions to the right? _____   (c)
Center the spot. Apply a small DC voltage to the horizontal (X) input and verify your answer experimentally. Position of AC-DC-GND switch? _____ Observed voltage _____.   (d,e)

---

*CH2 on some scopes.

(a)   .5 ms/div   (b)   The LEVEL is set above or below the peak of the triggering signal, so the sweep does not trigger.

2.88   Reset the LEVEL control so that the trace reappears. Now set the triggering mode to AUTO. What happens to the trace now when the LEVEL control is turned too far to either side of zero? _____

_____ (a)

Now reset the level control so that the trace becomes stable. Should changing the MODE switch back and forth between AUTO and NORMAL affect the level at which the sweep is triggered? _____ (b)

Verify this experimentally.

(a) decrease setting by a factor of 10 (to .2 volts/div), or on 922 rotate the VOLTS/DIV switch until 2 volts/div appear in the X10 probe window

(b) V_2 = 4.9, 4.9, 4.8, 3.8 (try again if your answers don't check within ±.2)

(c) The probe has a higher input impedance so it does not load the circuit as much.

THIS IS A GOOD PLACE TO TAKE A BREAK

Use of the Differential Mode

| 3.91 | When the differential mode of operation is selected, the CH1 and CH2 vertical amplifiers are linked together to form a differential amplifier. The CH1 input becomes the +INPUT and the CH2 input becomes the −INPUT. For proper differential operation, the CH1 and CH2 volts/div settings must be the same.

If the CH1 and CH2 inputs are v_1 and v_2, respectively, what should be the vertical deflection in the DIFF mode if S_v is .5 volts/div for both channels?

$$y = \underline{\hspace{4cm}}$$

In the rest of this part, the CH1 input will be referred to as the +INPUT and the CH2 input as the −INPUT.

Select the differential mode of operation on your scope. Depending on the scope, this is done by (a) depressing the DIFF vertical mode button, (b) selecting the CH1 − CH2 setting or (c) selecting the CH1 + CH2 (ADD) setting and pressing the INVERT button, which inverts CH2 to give CH1 + (−CH2) to obtain proper differential operation.

2213: Also set vertical mode to BOTH.

| 4.85 | Hint:

If we connect the same signal to both the vertical and horizontal inputs, what should we see if the phase shift in the vertical and horizontal amplifiers is the same? _____

What should we see if the phase shift is slightly different?

_____

Now go back and answer frame 4.84.

(a) approx. 3 to 5 volts are required for a full-scale deflection of 5 div.

(b) divide the measured voltage by 5 to get the sensitivity.

(c) +8 volts   (d) DC   (e) +8 volts

| 1.92 |
| 922 ONLY |

With no voltage applied to the horizontal input, set the horizontal sensitivity knob to X10 and recenter the spot. Now measure the horizontal sensitivity.

$$S_h = \underline{\hspace{4cm}} \tag{a}$$

CAL X-Y ONLY

Set the scope controls so that 5 volts will deflect the spot 3 divisions to the right of center.

Observed setting of horizontal VOLTS/DIV _____ (b)

Red gain control (calibrated/uncalibrated) _____ (c)

(a) The trace becomes unstable. The sweep waveform is no longer syncronized with the vertical input.

(b) NO (The AUTO position is just like the NORMal position if a triggering signal is present.)

| 2.89 |

For each of the following waveforms, set the SLOPE and LEVEL so that the trace begins at the leftmost graticule line with the slope and level shown.

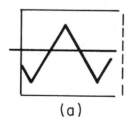
(a)
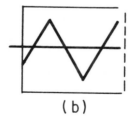
(b)
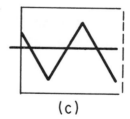
(c)
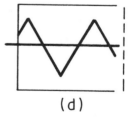
(d)

Record the SLOPE and LEVEL for each waveform:

	(a)	(b)	(c)	(d)
SLOPE				
LEVEL				

$$y = \frac{v_1 - v_2}{S_v} = 2(v_1 - v_2)$$

3.92 | Next, we will measure the CMRR for the differential amplifier. In preparation for this measurement, proceed as follows: Connect the output of the sine-wave oscillator to both the +INPUT (CH1) and the −INPUT (CH2) (in parallel). Use a frequency of 5 kHz and set the oscillator amplitude for 4 volts peak-to-peak. With the −INPUT grounded, adjust the scope so that about 4 cycles of the sine wave cover the entire screen.

Now set both the +INPUT and −INPUT to AC. Set the trigger mode so that you get a trace (even though the effective vertical input is 0.)

If the amplifier were ideal, the deflection would be _____.

Adjust the VOLTS/DIV (both channels the same) to obtain the largest deflection you can get without distorting the sine wave on the screen.

If the maximum peak-to-peak deflection is about 0.3 div or larger, turn to the next frame.

If the maximum peak-to-peak deflection is less than about 0.3 div, the CMRR of your scope is probably too large to measure by this method. In this case, we will intentionally mismatch the CH1 and CH2 amplifier gains. This will decrease the effective CMRR so that it can be measured.

Set CH1 and CH2 VOLTS/DIV to 0.5. Then turn the CH1 VARiable control to obtain 0.2 div peak-to-peak deflection.

Adjust the VOLTS/DIV (both channels the same) to obtain the largest deflection you can get without distorting the sine wave on the screen.

Answer to 4.84:
Connect the oscillator to both the vertical input and the horizontal input. Phase shift is the same if the trace is a diagonal line (/).

4.86 | Carry out the above procedure. Set the input switch(es) to DC. Adjust the scope and oscillator controls to obtain a trace which occupies nearly the full width and height of the screen. Observe the phase shift error as the frequency is varied from 20 Hz to 40 kHz.

Over what frequency range is the error negligible?

_____ to _____. At what frequency in   (a)

the range 20 Hz to 40 kHz is the error maximum? _____   (b)

(a) Your S_h measurement should be about $\frac{1}{10}$ the S_h measurement of frame 1.91.

(b) 1 volt/div   (c) uncalibrated

1.93 By now you should know how to connect and adjust the scope to deflect the spot by any desired amount in either direction.

922 ONLY
{ Set the horizontal sensitivity knob to the X1 position. With the spot initially centered, connect the DC power supply to the appropriate input terminals and adjust the power supply and scope controls so that the spot is deflected to the position x = 1, y = 2. (Do not readjust the position controls once the spot is centered and do not readjust the horizontal sensitivity knob.)   (a)

CAL X-Y ONLY
{ With the spot initially centered, connect 5 volts from the DC power supply to the appropriate input terminals and adjust the scope controls so that the spot is deflected to the position x = 4, y = 3. (Do <u>not</u> adjust the position controls once the spot is centered.)   (b)

(slope)   - + - +
(level)   - - + +

2.90 Up to this point you have been using the function generator as a signal source. Now you will use a sine wave oscillator (HP 204C or similar) instead. Examine the controls and terminals on your oscillator. If it is type HP 204C it will have three terminals at the bottom arranged as follows:

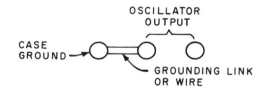

The left terminal is a case ground (which is connected to the building ground if there is a 3-wire line cord). The sine-wave signal is taken from the right two terminals. For normal operation, a grounding link or wire is connected between the center terminal and the case ground as shown.

Observe the oscillator output for several different frequencies and for several settings of the oscillator amplitude control.

Trigger mode should be set to AUTO.

The ideal deflection is 0.

3.93 | The observed peak-to-peak deflection is _____ div. with (a)

S_v = _____. (b)

Since the input signal is 4 volts peak-to-peak for both v_+ and v_-,

the common-mode gain is K_c = _____ div/volt (c)

The difference mode gain is $K_d = 1/S_v$ = _____ (d)

Therefore, the common-mode rejection ratio is CMRR = _____. (e)

(a) 20 Hz to about 10 kHz (this will vary from scope to scope but you should notice some phase shift error above 10 kHz).

(b) 40 kHz.

4.87 | At 20 Hz observe the effect of switching the <u>vertical</u> amplifier input from DC to AC. Explain. _____

_____ (a)

What is the minimum frequency at which the AC position can be used with-

out causing any noticeable phase shift error? _____ (b)

CAL
X-Y
ONLY

Now set both the vertical and horizontal inputs to AC and check the phase error at all frequencies in the range 20 Hz to 10 kHz

Is the error still negligible over the entire range? _____ (c)

You should have connected the + terminal of the power supply to both the vertical and horizontal scope inputs.

922: | If your x deflection isn't exactly 1 div, adjust the power supply voltage. Then if your y deflection isn't exactly 2 div, adjust the red VARIABLE knob on VOLTS/DIV.

CAL
X-Y: | With the CH1 and CH2 sensitivities set for 1 volt/div, you should have adjusted the associated red VARiable knobs to achieve the correct x and y deflections.

THIS IS A GOOD PLACE TO TAKE A BREAK

Scope Traces Due to Time-Varying Inputs

1.94 We will now study the effects of applying AC signals to the scope terminals. We will still be operating the scope with an external horizontal input connected to the horizontal (X) amplifier (sweep disabled) so that there will be no horizontal deflection unless we apply a horizontal input voltage. Check to see that the scope is set so that there is a small dot in the center of the screen when no inputs are connected.

2.91 Put a 200 Hz (922: use 400 Hz) sine wave into the vertical input. Use AC trigger coupling* and adjust the scope controls to obtain the following picture.

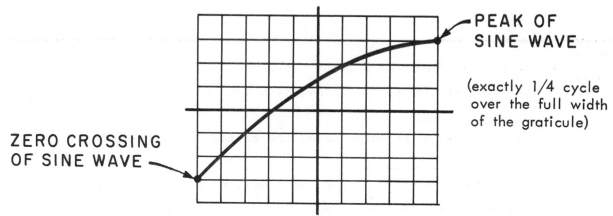

PEAK OF SINE WAVE

(exactly 1/4 cycle over the full width of the graticule)

ZERO CROSSING OF SINE WAVE

Suggestion: With one or two cycles of the sine wave on the screen, adjust the vertical position and oscillator output to give the desired zero line and peak deflection. Then adjust the SEC/DIV and triggering controls to give the desired trace.

---

*922: Trigger coupling is always AC, so no switch setting is necessary.

2213: Selecting INTernal triggering and AUTO mode automatically selects AC trigger coupling.

(a,b,d) Answer will vary from scope to scope

(c) Divide answer to (a) by 4.

(e) Divide K_d by K_c. The result should be above 50 with CH1 and CH2 calibrated. If you uncalibrated the CH1 gain in order to make the measurement, the measured CMRR should be about 40. (If you measured a value below 50 with both channels calibrated, your scope may need adjustment. Ask the supervisor to check.)

3.94 Now turn up the oscillator amplitude slowly. With VOLTS/DIV set to .2, at what input amplitude does distortion start to occur?

_____

(The input amplitude should be measured with the -INPUT grounded after distortion has been observed in the differential mode.)

(a) In the AC position, the blocking capacitor causes phase shift at low frequencies.

(b) about 200 to 500 Hz

(c) Yes

4.88 Now reconnect the phase shift network.

At what frequencies is the phase shift $+90^{\circ}$ or -90°? _____ ,

_____

Can you tell which is which using the ellipse method only?

_____   (a)

Since the phase shift of the network varies smoothly as the frequency varies, we can determine the sign of the phase shift from the dual trace data previously taken. From Table 4-1, at frequencies less than 1500, the sign of the phase shift is _____ and at frequencies   (b)

greater that 1700 the sign is _____.   (c)

At what frequency is the phase shift +90? _____   (d)

-90? _____   (e)

1.95 We will use a function generator (Tektronix FG 501 or similar) as a signal source. Study the function generator operating controls as shown in the instruction manual for your function generator. Is the outside of the BNC connector on your function generator output connected to

the generator chassis ground and line cord grounding pin? _____ (a)
Verify your answer using the continuity tester. When both the function generator and scope are plugged in to the AC line, will the grounds be

connected together? _____ What would happen if the (b)
center of the function generator output connector was connected to the

outside of a scope input connector? _____ (c)
Set the AC-DC-GND switch on the vertical (Y) input to the appropriate position for observation of very low frequency signals. To what position

did you set it? _____

(d)

Check your results as follows. Change the SEC/DIV switch setting to twice its present value. You should now see the waveform shown at the right (with the peak in the center).

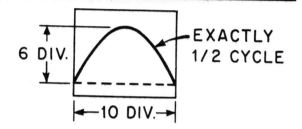

If you get this waveform, turn to the next frame.
If you failed to get the right waveform, turn back to frame 2.91 and try again. If you need help, continue with this frame.

2.92 Try using the following procedure:

(a) Ground the vertical inputs and obtain a horizontal line on the screen. Position this line to one division from the bottom of the grid.

(b) Observe one or two cycles of the 200 Hz (922: use 400 Hz) sine wave and set the peak amplitude for 6 divisions above the zero line established in (a). (Adjust the VOLTS/DIV or oscillator output or both as required.)

(c) Check the position of the zero line as in step (a) and repeat steps (a) and (b) if necessary.

(d) Set the trigger source to INT, trigger coupling to AC*, trigger slope to +, and adjust the trigger level until the sweep triggers at the zero crossing of the sine wave.

(e) Position the zero point to the left edge of the grid if necessary.

(f) Set SEC/DIV (including red variable knob) to give exactly one quarter cycle on the screen.

(g) Repeat (e) and (f) if necessary. When you are satisfied that the scope picture is exactly as shown in frame 2.91, check your answer by the method described above.

*922 ONLY: Switch setting not necessary.

2213: Use AUTO trigger.

Answer will be in the range 10 to 25 volts peak-to-peak for most scopes.

3.95 The trace is centered with no inputs applied, and both inputs are then set to DC. If a 6 volt peak-to-peak (3 volts peak) sine wave is then applied between the +INPUT and ground, +4 volts DC is applied between the −INPUT and ground, and S_v is set to 2 volts/div (calibrated), the vertical deflection should be

$$y = \text{\underline{\hspace{4cm}}} \qquad \text{(a)}$$

Sketch the expected trace.

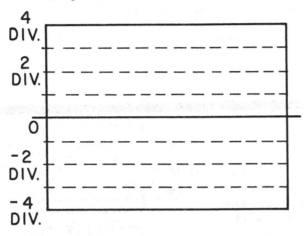

(b)

Verify your answer by observation.

(a) No. (b) − (c) +

(d) 3600 to 4100 Hz (e) 620 to 700 Hz

If your answers don't check, try again. When you have obtained answers which are within the prescribed tolerance, record them in the table.

4.89 Still using the ellipse method, measure the phase shift at 1000 Hz.

Adjust the vertical deflection for about 8 divisions peak-to-peak to obtain maximum accuracy.

$$|\theta| = \text{\underline{\hspace{4cm}}}$$

Determine the sign of θ from previous data.

$$\theta = \text{\underline{\hspace{4cm}}}$$

(a) yes   (b) yes

(c) the output would be shorted out through the scope ground

(d) If your answer is <u>AC</u>, continue with this frame; otherwise turn to the next page.

1.96 When the switch is in the AC position, there is a blocking capacitor in series with the input as shown in frame 1.71. When the frequency of the input signal is very low, the impedance of the blocking capacitor $(1/\omega C)$ will be very _____, and the input voltage to the scope will be _____.

Now turn back to 1.95 (d) and try again.

2.93 Connect a 250 Hz square wave to the vertical input and adjust the scope to display 5 cycles of the waveform. Now connect a sine-wave oscillator to the external trigger input (watch the grounds). Set the oscillator to 250 Hz and switch the trigger source to EXT. Turn up the oscillator amplitude about half way to assure that the triggering signal is adequate. Can a stable picture be obtained if the sine wave is not exactly the same frequency as the square wave? _____ (a)

Very carefully adjust the oscillator frequency to obtain a nearly stationary picture on the screen. Even with the most careful adjustment, the picture will gradually drift across the screen. What does this imply about the frequencies of the two oscillators? _____ (b)

<u>Compute</u> two other sine-wave oscillator frequencies less than 250 Hz at which a stationary picture could be obtained. (The vertical input and scope controls remain the same as above.) Computed values:

_____   _____ (c)

(a) 1/2 (3 sin ωt −4) =
        1.5 sin ωt − 2

(b)

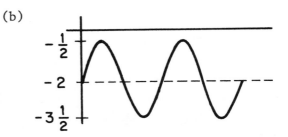

If your prediction and observation
don't agree, go back and try again.
(Check CH1 and CH2 VOLTS/DIV settings.)
Note that the deflection should be −2
div if you ground the +INPUT.

3.96 Measure the common-mode gain and compute the common-mode rejection

ratio using a 40 kHz sine wave input with 4 volts peak-to-peak am-

plitude. Note that the common-mode rejection ratio varies with the

setting of VOLTS/DIV.

K_c = _____ (a)

CMRR = _____ (b)

measured with S_v = _____ (c)

Did you center the ellipse horizontally before reading the deflections?
Your answer should check within 3° of the value measured by the dual
trace method. If it doesn't, find out why before continuing. When you
get a correct answer, record it in the table.

4.90 Carefully center the Webb mask over the graticule lines.

Adjust the trace so that with the vertical inputs grounded, the trace

lies along line _____ and with the horizontal input dis- (a)

connected the trace lies along line _____. (b)

Read θ from the Webb mask.   θ = _____ or _____

Which of these values is correct? _____ (c)

Answer to 1.95 (d): You should have set the switch to DC. (at very low frequencies, the AC position cannot be used because the impedance of the blocking capacitor is too high.)

1.97   With no horizontal input, if a sine wave is applied to the vertical input, describe the expected motion of the spot at very low frequencies.

_____ (a)

What should be observed at high frequencies?

_____ (b)

After writing your answers, observe the above using the following procedure:

   (a)   For this and the following frames use a vertical sensitivity of 1 volt/div and adjust the signal amplitudes as required for a reasonable deflection.

   (b)   Connect the function generator to the vertical input. Use sine wave output and vary the frequency between about 0.1 and 100 Hz. (If your function generator has a DC offset control, make sure that it is turned off or set to zero.)

   (a)   No.

   (b)   they are not exactly the same; they are not synchronized; they are not constant so that one is changing with respect to the other

   (c)   Verify your own answers experimentally. (Since the oscillator dial reading is not exact, a slight adjustment of the oscillator frequency will probably be necessary to get a stationary picture.)

2.94   (a)   Now connect the same sine-wave oscillator to both the vertical input and to the external trigger input. Set the VOLTS/DIV to .1, the TRIGGER LEVEL to 0, and the mode to NORMal. Switch the trigger source back and forth between EXT and INT. Notice that you get a stationary picture in either position since the triggering voltage comes from the same oscillator in either position.

   (b)   With trigger source set to EXT, increase the VOLTS/DIV until the sine wave goes to zero and only a horizontal line remains.

   (c)   Reset VOLTS/DIV to obtain the original sine wave again. Switch trigger source to INT and increase VOLTS/DIV. Observe that the whole trace disappears when the sine wave amplitude gets small.

   (d)   Explain why a horizontal line remains in (b) but the trace completely disappears in (c).

         _____

   _____

   _____

   (If in doubt, study Fig. 5-3 on p. 234 again.)

You should have set the oscillator output to 4 volts peak-to-peak (with the -INPUT grounded). Then with the oscillator connected to both +INPUT and -INPUT, you should have set S_v for an adequate deflection. If this deflection is y_1 (peak-to-peak), then $K_c = y_1/4$, $K_d = 1/S_v$, and CMRR = K_d/K_c. CMRR should be greater than 100.

3.97 This exercise will test your ability to operate the scope.

(a) Rotate the following controls fully counterclockwise: focus, intensity, vert. position, horiz. position, VOLTS/DIV and variable, SEC/DIV and variable. Change all switch and pushbutton settings from their present settings.

Keep a record of the time required to do parts (b) and (c)

(b) Check the calibration of both vertical amplifiers. Check the probe and adjust it if necessary.

(c) Using the probe, display 3 cycles of a 2 volt peak-to-peak 400 Hz sine wave on the scope with the waveform starting at this point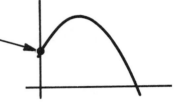

How long did it take you to do parts (b) and (c)?

_____

(a) B - B'   (b) A - A'

(c) Again, your answer should check within 3° of the value measured by triggered sweep. Make sure your ellipse is tangent to all four sides of the square on the Webb mask.

4.91 Now measure the phase shift at 100, 300, 3000, and 10,000 Hz. In each case, first compute the angle from distances measured on the ellipse and then read the value from the Webb mask.

Does the amplitude of the phase shift network output (vertical) change with frequency? _____ (a)

If it does change, what adjustment must be made before reading the phase angle from the Webb mask? _____

_____ (b)

Record the computed angles and the angles read from the Webb mask in Table 4-1. (The table should now be complete except for (c) the shaded areas.)

(a) The spot will move up and down (at the sine wave frequency).

(b) The moving spot will become a vertical line. (If your answer was "a sine wave" you are wrong because there is no horizontal signal and hence no horizontal deflection.)

1.98 If the sine wave is changed to a triangular wave, should there be any

appreciable change in the observed picture? _____ (a)

After writing your answer, observe this on the scope.

If the triangular wave is changed to a square wave, what should be

observed at very low frequencies (assume that the square wave is ideal

so that it jumps <u>instantaneously</u> between its maximum and minimum

values)? _____

_____ (b)

At high frequencies? _____ (c)

After writing your answers, observe the above phenomena on the scope.

(d) In (b) the triggering voltage remains constant, but in (c) the triggering voltage decreases as the vertical amplifier output decreases.

2.95 Now disconnect the external trigger lead and use only the internal trigger for this part.

(a) With VOLTS/DIV set to 10 and trigger mode set to AUTO, obtain a sine wave on the screen. Gradually turn down the oscillator output until the sine wave goes to zero and only a horizontal line remains.

(b) Reset the oscillator output to obtain the original sine wave and set trigger LEVEL to 0 and mode to NORMal. Now turn down the oscillator amplitude and observe that the whole trace disappears when the amplitude gets small.

(c) Explain why a horizontal line remains in (a) but the trace completely disappears in (b).

_____

_____

_____

_____

Hum and Noise Pickup

3.98 The leads which connect the scope (or other measuring instrument) to the signal being observed may pick up extraneous signals. Such extraneous signals may cause fuzzy pictures on the screen and make accurate measurements impossible. In this section we will observe some of the conditions under which extraneous signals can be picked up, and we will investigate some things which can be done to eliminate them.

Attach an unshielded long lead (about 30" long) to the CH1 vertical amplifier input* and set SEC/DIV to 5 ms. Leave the lead stretched out on the bench with the other end not connected. Adjust the scope to obtain a good picture. What is the frequency of the signal which is picked up by the lead? _____   (a)

Observe that the trace will synchronize on LINE trigger as well as INT trigger.

What do you think is the source of the signal being picked up?

_____   (b)

_____

*Attach a BNC to banana plug adaptor to the CH1 input. If such an adaptor is unavailable, attach the long lead to the red clip on the coaxial cable lead.

(a) Probably not significantly

(b) Adjust oscillator output and/or oscilloscope sensitivity so ellipse is tangent to all four sides of the square. (This is most easily accomplished by grounding one input at a time as in frame 4.90).

(c) Computed angles should agree with values from Webb mask within 3°.

4.92 Now measure the phase shift of v_4 with respect to v_1 at 5000 Hz when the network is connected as follows:

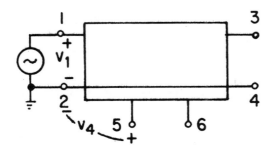

θ (calculated from ellipse) =

$\sin^{-1}$ _____ = _____   (a)

θ (read from Webb mask) =

_____   (b)

Check your answers by the dual trace method.

θ (by dual trace) =

_____   (c)

(a) No. (except the spot will now travel up and down at a uniform rate)

(b) The dot will jump up and down (between the two positions which correspond to the maximum and minimum of the square wave.)

(c) The dot will appear in both places simultaneously (because of persistence of the screen.)

1.99 | Now ground the vertical input.

[CAL X-Y ONLY: Set the horizontal (X) AC-DC-GND switch to the appropriate position for very low frequency AC signals. Use 1 volt/div horizontal sensitivity.]

If a sine wave is applied to the horizontal (X) input, what should be observed at low frequencies? _____ (a)

at high frequencies? _____ (b)

After writing your answers, observe the above on the scope.

For AUTO trigger the sweep triggers even when no triggering signal is present. In (b) the trace disappears when the triggering signal becomes too small.

2.96 | A 6 volt peak-to- peak, 1000 Hz sine wave is added to a 4 volt DC level to give

$$v(t) = 3 \sin 2000\pi t + 4$$

If v(t) is observed with a scope set to 2 volts/div, AUTO trigger, and LEVEL = 0, sketch the expected waveform if (a) the vertical AC input is used and (b) the vertical DC input is used.

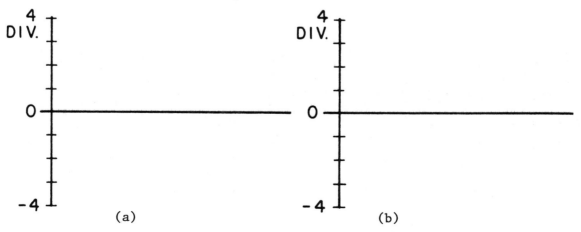

(a)　　　　　(b)

(a)   60 Hz   (b)   the 60 Hz AC line

3.99 An adaptor plug is available which converts the three-terminal grounding plug on the scope line cord to a two-terminal plug without ground.
Attach such an adaptor plug to the line plug on the scope. When the
adaptor plug is in place, the ground terminals on the scope panel
(are/are not) _____ connected to the building   (a)
ground.

With the adaptor plug in place, is the 60 Hz pickup more or less
than when the scope is grounded to the building ground? _____ (b)

Leave the adaptor plug connected until you have completed frame 3.104. It
will make it easier to observe the effects of hum and noise pickup.

Your answers should agree within 3°. If they don't, go back and find out
why.

Before continuing, have your lab instructor or supervisor check your
answers and sign here: _____

4.93 If you vary the frequency continuously and observe the ellipse, how
can you tell when the phase angle goes through 0° or 180°?

_____ (a)

The phase shift of most networks is a smooth function of frequency;
therefore, when the phase angle passes through zero degrees, a plot
of phase angle vs frequency will look like

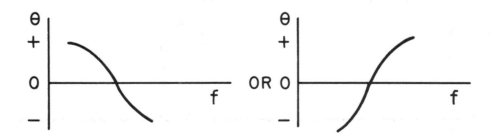

By watching the ellipse, how can you tell when the phase angle changes
sign as the frequency is varied?

_____ (b)

(a) The spot will move back and forth at the sine wave frequency.

(b) The moving spot will become a horizontal line.

1.100 With the sine wave still applied to the horizontal input, what should be the effect of applying + 2 volts DC to the vertical input?

_____

After writing your answer, observe this on the scope.

Verify your answers experimentally using the procedure below.

2.97 Set the sine-wave oscillator to 6 volts peak-to-peak, and set the DC power supply to 4 volts. To add the DC voltage to the sine-wave voltage, you must connect the DC power supply in (series/parallel)

_____ with the oscillator. Indicate the   (a)
proper connections to the scope on the diagram below. Make sure that
you have a common ground wire between the oscillator and scope and that
the power supply polarity is correct.

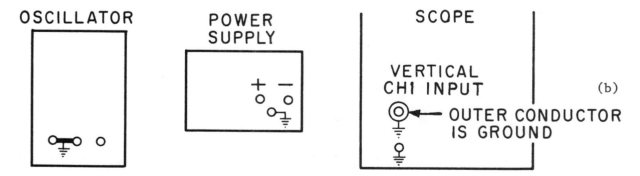

Connect the circuit and verify your answers to the preceding frame.
Make sure the trace is centered when the scope input is grounded.
When the input switch is changed from DC to AC, the trace moves

(up/down) _____ _____ div.   (c)

(a)   are not   (b)   more (If you observed less or no change, your scope
                            might have a bad plug on the line cord; ask the
                            supervisor to check.)

3.100   Electric and magnetic fields of line frequency are always present in the
        lab (or anywhere in the vicinity of AC power lines, motors, transformers,
        and AC operated equipment). These stray fields will induce unwanted
        signals in our leads if we do not take the proper precautions. The magni-
        tude of the signal picked up by a circuit will depend on its impedance.

        Connect a second long lead to the scope ground terminal.
        Stretch out the leads to nearly full length and connect them to a one
        megohm resistor mounted on a circuit board. Observe the amplitude of
        the 60 Hz hum signal which is displayed on the screen. Repeat with the
        resistance changed to 100 kilohms, 10 kilohms, and 1 kilohm:

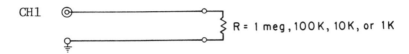

        The amplitude of the hum signal which is picked up is largest for
        _____ (high/low) impedance circuits.

(a)   ellipse becomes a straight line

(b)   ellipse closes to a straight line and then opens again

4.94   You need to measure the phase shift of a network over a range of frequencies
        from 100 Hz to 100 kHz. You know that the sign of the phase angle is
        positive at low frequencies, so you decide to use the ellipse method. As
        you vary the frequency from 100 Hz to 100 kHz, the phase angle varies con-
        tinuously and you observe the following sequence:

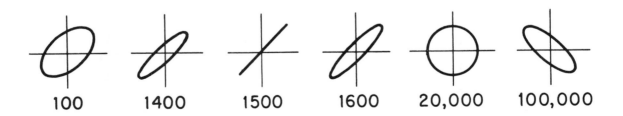

        The ellipse closes only at 1500 Hz as shown. Over what frequency range,
        if any, is the phase angle negative? _____

The trace will move up 2 divisions.

1.101 What should be observed if the same sine wave is applied to both horizontal and vertical inputs? _____

After writing your answer, observe this on the scope.

(a) series (b) right terminal of oscillator to power supply −, + terminal of power supply to center of scope vertical input

(c) down 2

Answers to 2.96: (a) (b)

If your observed waveforms are not the same as above, go back and try again.

Trigger Coupling

2.98 We will now investigate the effect of DC trigger coupling.

922 ONLY: This scope allows only AC trigger coupling. Turn to frame 2.101.

CAL X–Y ONLY

2213 ONLY: When INTernal trigger is used, the trigger coupling is selected by the AUTO–NORM mode switch. When AUTO is selected, the trigger coupling is AC, and when NORM is selected, the trigger coupling is DC. In this and the following frames, use the AUTO–NORM switch to switch between AC and DC trigger coupling.

With the same input as in 2.96, vertical input set to DC, triggering LEVEL set to 0 or slightly −, triggering mode set to NORMal, determine the following:

(a) Will the sweep trigger if DC trigger COUPLING is used?

_____ (a)

(b) Will the sweep trigger if AC trigger COUPLING is used?

_____ (b)

Explain the difference:

_____

_____ (c)

3.101 Reconnect the 1 megohm resistor to the long leads.

Connect a second 1 megohm resistor to the CH2 input and ground with short leads (about 6 or 8 inches). Switch back and forth between CH1 and CH2 and compare the relative amplitudes of the signals.

Hum pickup can be reduced by using _____ _____.

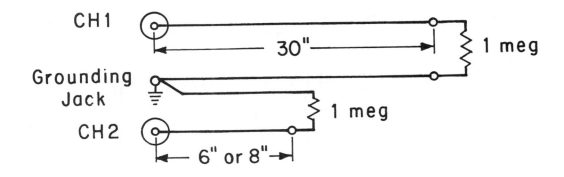

---

f > 1500 Hz

4.95 The phase shift of a network is to be measured over a wide range of frequencies. Each time the frequency is changed the time axis must be recalibrated if you are using the _____ or   (a)
_____ method.

However, it is unnecessary to calibrate either axis precisely when using the _____ method.   (b)

If you wish to make a series of phase measurements over a wide range of frequencies without having to change the circuit connections, you should use the _____ method.   (c)

If the sign of the phase angle is not known and you wish to determine it, you should use the _____ or _____ method.   (d)

a diagonal line (since the vertical and horizontal deflections are about equal)

1.102 For the following circuit, if $v_h = V_h \cos \omega t$, then $v_v \simeq V_h/\omega RC \sin \omega t$ as was derived in the preparation. What should be observed on the scope? _____

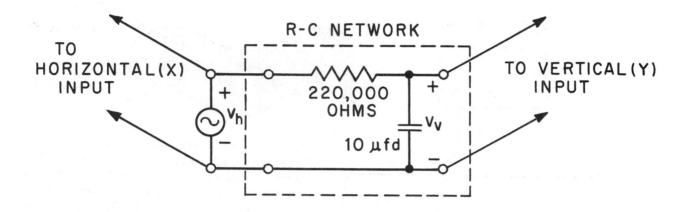

(a) no   (b) yes

(c) For DC coupling, the entire signal is used for the trigger input, and since this signal never goes to 0 or negative, triggering never occurs. For AC coupling, only the AC component is used so the trigger circuit input does go through 0 and triggering occurs.

2.99 Select AC trigger coupling and observe the effect of slowly varying the DC supply voltage. Then repeat using DC trigger coupling.

Note that in one case, the waveform moves up and down, but the triggering point on the waveform remains unchanged.

CAL
X-Y
ONLY Note that in the other case, the trigger point on the scope screen does not change, so that the waveform moves sideways as well as up and down.

Explain the difference: _____

_____

_____

_____

short leads.

3.102 Disconnect the short leads from CH2. Connect a long lead to the ungrounded output terminal of the sine-wave oscillator and stretch it out across the leads connected to CH1 of the scope (see diagram below). Set the oscillator to 10 kHz, turn the oscillator amplitude all the way up, and observe the signal which is picked up by the scope leads. The observed signal will consist of a 10 kHz signal with some 60 Hz added to it. It should look like this:

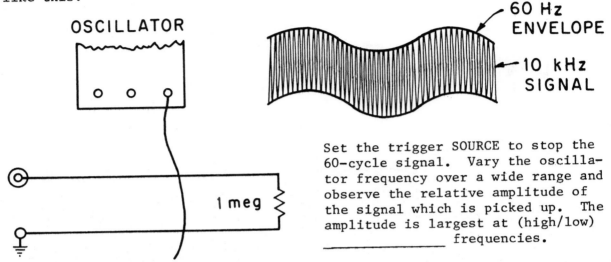

Set the trigger SOURCE to stop the 60-cycle signal. Vary the oscillator frequency over a wide range and observe the relative amplitude of the signal which is picked up. The amplitude is largest at (high/low) _____ frequencies.

(a) dual trace or trig. sweep   (b) ellipse   (c) ellipse

(d) dual trace or trig. sweep

4.96 On the figure below indicate the connections to the scope to measure the phase shift of v_3 relative to v_1 using the ellipse method. (Note that there is not a common ground).

922
ONLY

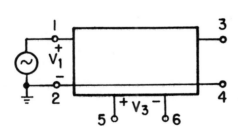 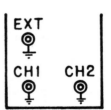

An ellipse (or a circle if the peak X and Y deflections are equal)

1.103   Obtain an R-C network (R = 220,000 ohms, C = 10 microfarads) from the supervisor and connect the circuit as shown in frame 1.102 using the function generator as a sine-wave source. Turn up the sine-wave amplitude, set the frequency to 1 Hz, and make the necessary adjustments so that the spot travels in an approximately circular path. (Adjustment of the vertical gain, both calibrated and uncalibrated as well as the position may be necessary.)

[CAL X-Y ONLY: Also adjust the horizontal gain as necessary.]

If you have trouble getting your circuit to function properly, turn to frame 1.104.

Which direction does the spot travel around the circle?

_____

Turn to frame 1.105.

AC trigger coupling: DC component does not affect the trigger point, so triggering always occurs at the same point on the AC waveform.

DC trigger coupling: Total value (AC + DC) determines the trigger point. Since trigger level is constant, triggering occurs at the same distance above the x-axis.

2.100   With DC trigger coupling, adjust the DC power supply and the trigger controls to obtain the following picture (with the zero line still at the center of the screen):

CAL
X-Y
ONLY

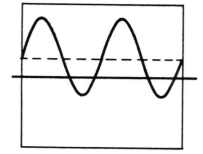
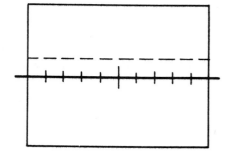

Sketch the waveform you would expect to see if the DC component of the input voltage is set to 0. Verify your answer by observation.

3.103 Reset the oscillator frequency to 10 kHz. Connect a pair of long leads to the CH2 input and ground terminal. Twist this pair of leads together and connect them to a one megohm resistor. Mount this resistor on the circuit board next to the other one (see diagram).

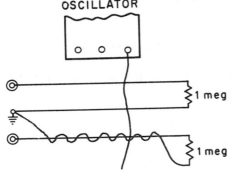

OSCILLATOR

1 meg

1 meg

Compare the relative amplitudes of the signals picked up by switching back and forth between CH1 and CH2. This illustrates that pickup can be reduced by using _____

_____. (a)

Replace the untwisted leads with a shielded (coaxial cable) lead connected from the CH1 input to the resistor. Is the pickup less with the shielded lead or the twisted leads? _____ (b)

v_1+ to external horizontal (x) input; v_3+ to +INPUT (CH1), v_3- to –INPUT (CH2); oscillator ground to scope ground

4.97 Measure the phase shift of v_3 relative to v_1 by the ellipse method for $f = 5000$ Hz.

922
ONLY

$|\theta| =$ _____ = _____ (a)

At what frequency (if any) is the phase shift $0°$? _____ (b)

$180°$? _____ (c)

Check your answer to (a) by the triggered sweep method.

$\theta =$ _____ (d)

1.104 (a) Check your circuit to make sure that it is connected as shown in the diagram below. Pay particular attention to the grounds.

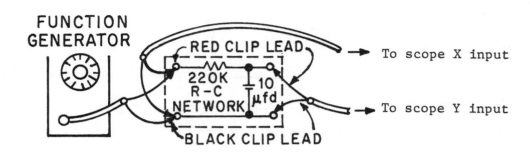

(b) [922 ONLY: X input is the External Horizontal (X) input, Y input is CH1]

[CAL X-Y ONLY: X input is CH1, Y input is CH2. Set horizontal (X) sensitivity to 1 volt/div.]

Adjust function generator output amplitude to give desired horizontal deflection on screen.

(c) If the trace moves off the screen when you adjust the vertical VOLTS/DIV switch and it cannot be brought back with the BEAM FINDER or position controls, check the function generator for a DC offset.

Now turn back to frame 1.103.

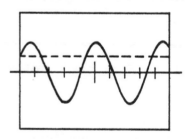

Note that the sweep still triggers at the same overall level but has changed relative to the AC component.

If you had difficulty with frames 2.99 and 2.100, review frames 2.40-2.45 before continuing.

2.101 (a) Set the scope to display exactly 4 cycles of a 500 Hz triangular wave (with no DC component).

(b) Now switch the trigger source to LINE. Can a stable picture be obtained? _____

(c) Explain. _____

(d) Suppose that the INT position of the SOURCE switch was defective and could not be used. With the vertical input the same as in (a), how would you set up the scope so that a stable picture could be obtained?

_____

_____

(e) Verify your answer to (d) experimentally.

(a) twisted leads   (b) shielded lead

3.104 State the effect of each of the following on pickup of stray signals:

circuit impedance

_____ (a)

signal frequency

_____ (b)

lead length

_____ (c)

type of leads

_____

_____ (d)

(a,d) -127° answers should check within 3°

(b) 920 Hz ± 30   (c) approximately 40 kHz

4.98 Given a choice between the ellipse method and the dual trace method, which one would you use

(a) to find the frequency at which the phase angle is 0° or 180° _____

(b) if you want to get a rough idea of how the phase angle varies when the frequency is varied over a wide range? _____

(c) if you must find the sign of the phase angle? _____

(d) to find the frequency at which the magnitude of the phase angle is 90°? _____

Counterclockwise (compare this with the answer obtained in the preparation, frame 1.32)

1.105 As derived in the preparation, if $v_h = V_h \cos \omega t$, then $v_v = (V_h/\omega RC) \sin \omega t$. Explain why the vertical sensitivity must be changed in order to maintain a circular trace as the frequency is increased. _____

Now gradually increase the frequency from 1 Hz to 10 Hz and observe the effect on the trace. As you increase the frequency, make the necessary adjustments to maintain an approximately circular trace.

(If the trace moves off the screen when you adjust the vertical VOLTS/DIV switch and it cannot be brought back with the BEAM FINDER or position controls, check the function generator for a DC offset.)

(a) Note that the sweep must be uncalibrated to obtain the required trace.

(b) No.

(c) 500 Hz is not a multiple of 60 Hz, so the sweep will trigger at a different point in each cycle (i.e., the vertical input and sweep are not synchronized).

(d) Set the SOURCE to EXT and connect the external trigger input in parallel with the vertical input.

(e) If you didn't get a stable picture, try again.

THIS IS A GOOD PLACE TO TAKE A BREAK

Using Dual Trace

2.102 We shall now use the dual trace feature of the oscilloscope to display two different waveforms simultaneously.

(a) Set the scope controls to display a horizontal line when the CH1 input is grounded.

(b) Center the CH1 horizontal line in the top half of the scope screen.

(c) Now set the scope controls to display a horizontal line when the CH2 input is grounded.

(d) Center the CH2 horizontal line in the bottom half of your scope screen.

(e) Set your scope to display a dual trace. (Set the appropriate vertical mode button or switch to DUAL, BOTH, or A&B).

(f) If your scope has an alternate-chopped switch, select ALTernate.

(g) Sketch the display on the scope screen to the right.

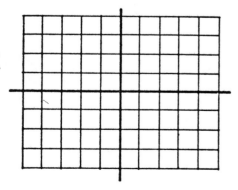

(a) the lower the impedance, the less the pickup
(b) the higher the frequency, the more the pickup
(c) the shorter the leads, the less the pickup
(d) straight leads pick up most, twisted leads somewhat
     less, and shielded leads the least.

Remove the adaptor plug from the line cord.

Measurement of Rise Time

3.105 Figure 3.105(a) shows an ideal square pulse. In practice, the voltage will
not instantaneously change from 0 to its maximum value, but the voltage will
rise more slowly as shown in Fig. 3.105(b). The time required for the voltage to change from 10% of its maximum value to 90% of its maximum value is
often referred to as the rise time of the pulse. For the pulse shown in

Fig. 3.105(b), the rise time is _____.

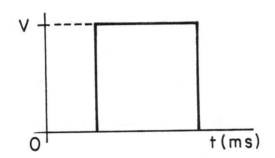
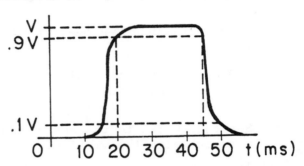

FIG. 3.105(a). IDEAL SQUARE PULSE   FIG. 3.105(b). NON-IDEAL SQUARE PULSE

(a) ellipse   (b) ellipse   (c) dual trace

(d) ellipse

THIS IS A GOOD PLACE TO TAKE A BREAK

Lissajous Figures

4.99 Show the connections required to display a Lissajous figure on the scope.

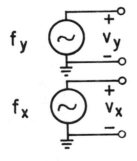
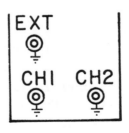

(a)

The scope should be set to the (sweep, X-Y) _____ mode.   (b)

The number of positive maximums on the Lissajous figure should be

determined when it is fully _____.   (c)

214

v_v decreases as frequency is increased (so S_v must be changed to compensate)

TURN TO FRAME 1.106

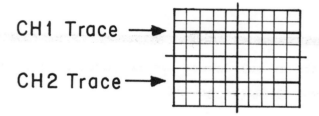

2.103 Obtain a 2200 ohm resistor and a 0.1 microfarad capacitor and connect the circuit shown below. (Be careful with the grounds.) Use a function generator to supply a 1000 Hz sine wave to the circuit, and set the output amplitude to the middle of the range. Set CH1 and CH2 vertical sensitivities to S_v = 2 volts/div.

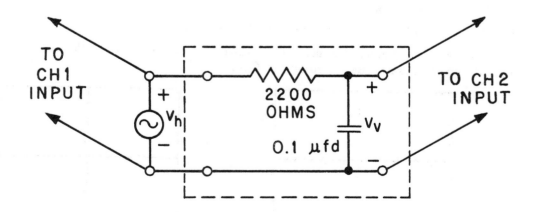

3.106 For practice in making rise time measurements we will use the circuit shown below to generate a series of non-ideal square pulses. Connect the circuit and display 2 or 3 of the output pulses on the scope. (Make sure the AC-DC-GND switch is properly set.)

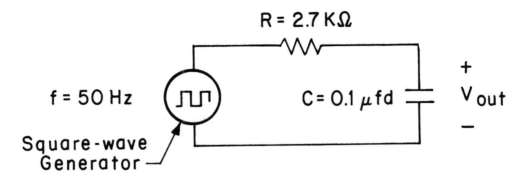

R = 2.7 KΩ

f = 50 Hz

C = 0.1 μfd

+ V_{out} −

Square-wave Generator

(a) v_x+ to external horizontal (x) input; v_y+ to vertical (y) input; both oscillator grounds to scope ground

(b) X-Y   (c) open

4.100 Connect two sine wave oscillators to your scope so that you can display Lissajous figures. Use the sine wave output of your function generator for the vertical input, and an additional sine wave oscillator for the horizontal input. Set the horizontal input to 200 Hz and the vertical input to 600 Hz. Adjust the oscillator amplitudes and your scope controls to obtain a Lissajous figure which nearly fills the screen. Use the fine frequency adjustment control* on your vertical input oscillator to stabilize the figure. Observe that the figure slowly changes from fully open to fully closed and back again.

Sketch the fully open figure in (a) below.

Now set the vertical input oscillator to 100 Hz and stabilize the figure. Sketch this Lissajous figure in (b).

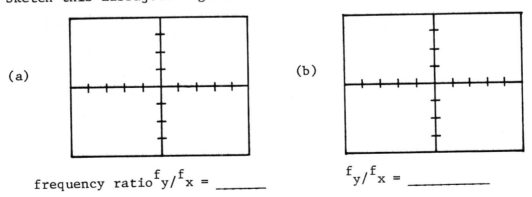

(a)

frequency ratio f_y/f_x = _____

(b)

f_y/f_x = _____

---

*Labeled frequency vernier on FG501

TURN TO FRAME 1.106

2.104 In the dual-trace mode, the internal trigger source can come from either CH1 or CH2.

922 ONLY: When the dual-trace mode is selected, CH1 is automatically selected as the trigger source. To use CH2 as the trigger source, you must press both the dual-trace and CH2 buttons.

CAL X-Y ONLY: The trigger source switch or buttons allow selection of CH1 or CH2 as the internal trigger source.

ALL: Set the appropriate scope controls to trigger off the CH1 signal in the dual trace mode. Set LEVEL to approximately 0 and the SLOPE to +. Adjust the oscillator amplitude so that the amplitude of the CH1 waveform is exactly 8 volts peak to peak. Now set the SEC/DIV to display exactly two cycles of the CH1 waveform. Adjust the LEVEL so that the scope triggers exactly when the triggering voltage is 0 volts. Sketch the scope display on the screen below.

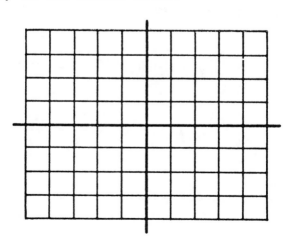

Scope trace should
be similar to:

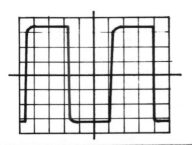

3.107 For convenience in making the rise time measurement, we will adjust the
position of the trace so that the 10% and 90% points lie on graticule
lines. If we position the trace as shown below, indicate where the 10%
and 90% points will be.

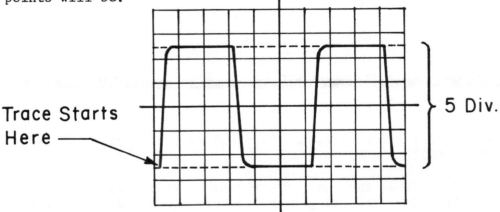

Trace Starts
Here

5 Div.

Make the necessary adjustments so that the trace is positioned as shown
above. Adjust the trigger line so that the trace starts as close to the
base line as practical.

Note: Some scopes have dotted lines on the graticule to facilitate this
adjustment. If you don't have dotted lines, first set the waveform
for 5 divisions and then shift it to the position shown.

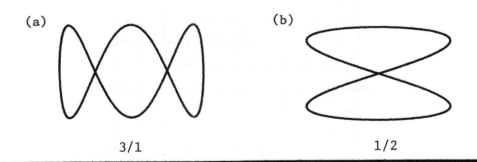

(a)   (b)

3/1   1/2

TURN TO FRAME 4.101

1.106 For each of the following scope traces, the vertical input is
$v_v = 2 \cos \omega t$.

The horizontal sensitivity is 1 volt/div.

In each case calculate the horizontal input voltage and vertical sensitivity.

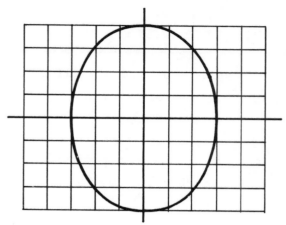
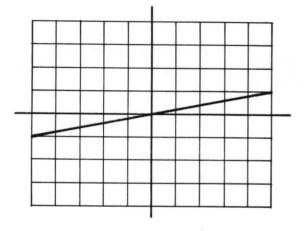

(a) $v_h(t) =$ _____

(b) $S_v =$ _____

(c) $v_h(t) =$ _____

(d) $S_v =$ _____

Check your answers by considering limiting values before you turn the page.
(Note that answers to (a) and (c) should be time functions.)

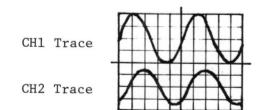

CH1 Trace

CH2 Trace

2.105 Without changing anything else, switch the internal trigger source from CH1 to CH2 and back several times. Why do both waveforms shift when CH2 is selected? _____

_____

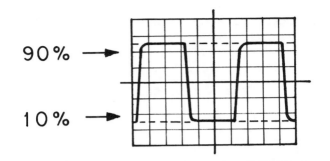

90% →

10% →

3.108 With the scope vertical setting left the same as in the preceding frame, what should be done so that the rise time can be read more accurately from the scope screen?

_____ (a)

Take the above action, adjusting the scope so that the rise time can be read as accurately as possible. Setting of SEC/DIV _____ (b)

Measured value of rise-time _____ (c)

4.101 Using the same horizontal frequency, adjust the vertical frequency to obtain Fig. (a) below. Observed vertical frequency setting. _____

Now change the horizontal frequency to obtain Fig. (b). Observed horizontal frequency. _____

(a)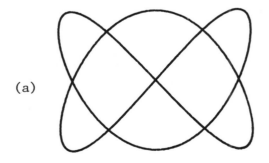

(b)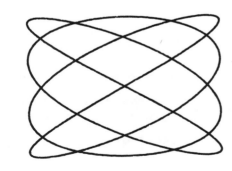

frequency ratio from
oscillator dial
f_y/f_x = _____

frequency ratio from
oscillator dial
f_y/f_x = _____

Do these frequency ratios agree with those obtained from the Lissajous figures? _____

(a) v_h = 3 sin ωt   (c) v_h = 5 cos ωt

(b) S_v = .5 volts/div   (d) S_v = 2 volts/div

1.107 You have completed Part I. You should now understand the relation between the voltages applied to the vertical and horizontal inputs of the scope and the resulting traces on the screen. Do Preparation Part II before continuing with the lab work.

For both waveforms, the sweep is triggered when CH2 goes through 0 with + slope.

2.106 Many scopes will allow manual selection of either the ALTernate or CHOP dual trace modes. For "fast" sweep rates (.5 msec/div or faster) it is generally best to use the _____ mode; whereas, the   (a) _____ mode is better for slow sweep rates. (Review   (b) frame 2.61 if you need help.)

922 ONLY { The 922 automatically determines whether to use CHOP or ALTERNATE when in the DUAL TRACE mode. For SEC/DIV settings of .5 ms/div and faster, ALTERNATE is used. For SEC/DIV settings of 1 ms/div and slower, CHOP is used. In the previous frame, the 922 used the _____ dual trace mode.   (c)

Remove the R-C network and ground the CH1 and CH2 inputs with the AC-DC-GND switches. Set the SEC/DIV switch to .2 ms/div. With this setting, you should use the _____ dual trace mode*. Now con-   (d) nect a 10 Hz square wave to the external trigger (X) input and set the scope controls and square wave amplitude to allow external triggering.

Describe and explain the scope display. _____

_____

_____   (e)

Now increase the frequency of the square wave until both traces are solid and visible at the same time. At about what frequency does this

occur? _____   (f)

---

*922 ONLY: This mode is automatically selected.

(a) spread out the waveform in the horizontal direction (decrease SEC/DIV)

(b) adjust the scope to obtain the following picture:

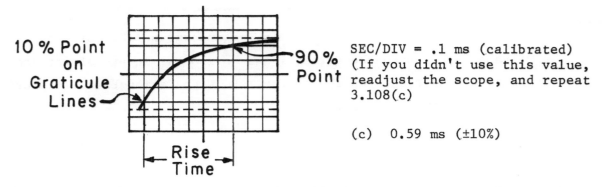

10 % Point on Graticule Lines

90 % Point

Rise Time

SEC/DIV = .1 ms (calibrated)
(If you didn't use this value, readjust the scope, and repeat 3.108(c)

(c) 0.59 ms (±10%)

3.109   Summary of procedure for measuring rise time:

1. Display 2 or 3 cycles of the pulse waveform on the scope and adjust the vertical sensitivity and position so that the peak-to-peak deflection is 5 divisions and the waveform is centered vertically.

2. Adjust the SEC/DIV so that the rising edge of the pulse occupies most of the screen width. Adjust the horizontal position so that the trace crosses the 10% point at the intersection point of two graticule lines.

3. Read the horizontal distance between the 10% point and the 90% point and compute the rise time.

(a) 3/2   (b) 3/4

Yes (within the calibration accuracy of the oscillator dials)

4.102   Set the horizontal oscillator to 300 Hz. Obtain stable Lissajous figures (with five or less vertical or horizontal maximums) for four different settings of the vertical frequency in the range 200 Hz through 500 Hz.

Vertical Frequency	Frequency Ratio

(a) ALTERNATE   (b) CHOP

(c) ALTERNATE   (d) ALTERNATE

(e) The scope displays first one of the traces, then displays the other, then displays the first again, etc. If the sweep is triggered at a slow enough rate, by the time the scope displays the second trace, the first trace has faded from the screen.

(f) 20 to 80 Hz

2.107

922 ONLY

Set the SEC/DIV to 1 ms/div. For this setting, the scope will use the _____ dual trace mode. Rotate the X1–X10 knob to the X10 position.   (a)

CAL X-Y ONLY

Set the SEC/DIV to 1 ms/div and select the CHOP mode. Select the X10 (or X5) sweep magnifier by pulling the appropriate knob or pushing the appropriate button.

Using internal triggering (level = 0, slope = +, S_v = 2 volts/div), display an 8 volt peak-to-peak 5000 Hz sine wave on both the CH1 and CH2 traces. Sketch the scope display on the screen below.

Why are the chop marks present? _____

_____ (b)

CAL X-Y ONLY

Do the chop marks disappear when ALTernate is selected? _____ (c)

3.110 Change the resistor in your network to 5600 ohms. Increase the amplitude of the square wave generator output.

Readjust the scope and measure the rise time as accurately as you can.

Setting of SEC/DIV _____(a) measured rise time _____(b)

Possible vertical frequencies: 200, 225, 240, 300, 375, 400, 450, 500
(Check your frequency ratios to see that they represent f_y/f_x.)

4.103 You have now completed Lab Part IV. You should know how to measure phase shift accurately by the dual trace method, the triggered sweep method, the ellipse method, and by using the Webb mask. You should also be able to determine frequency ratios from stable Lissajous figures.

Part V consists of a review and summary of the material covered in the first four parts. You may wish to study Part V as a review before you take the practical laboratory examination on use of the oscilloscope.

(a)   CHOP

(b)   The scope is not being triggered at a high enough frequency to produce
      enough overlapping CH1 and CH2 traces to eliminate the chopped lines
      (see frame 2.60.)

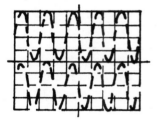

      The chopped lines may appear to be moving on your scope.

(c)   Yes

2.108   This completes Part II. You should now understand how to use the triggering
        controls to display periodic waveforms on the scope.

(a) 0.2 ms/div.   (b) 1.2 ms.

3.111 Look at your answer to frame 3.97. If your answer is more than 20 minutes you need more practice, so go back and rework frame 3.97.

This completes Lab Part III. You should be able to check the scope calibration and adjust the probe so that the scope can be used to make accurate measurements of voltage or time. You should now understand the functions of all of the standard scope controls.

TABLE 4-1

4.104 Phase shift for active network number _____
Measured using oscillator number _____

FREQUENCY (Hz)	θ MEASURED BY DUAL TRACE	θ MEASURED BY ELLIPSE METHOD	θ READ FROM WEBB MASK
100			
300			
	/////////	−90	/////////
1000			
	/////////	±180	/////////
3000			
	/////////	+90	/////////
10000			

PART V   REVIEW AND SUMMARY

This part consists mainly of a summary of the material presented in the first four parts arranged for more convenient reference. The appropriate parts of the programmed instruction in Parts I through IV should be studied before reading this review.

5.1   Functions of the Scope Controls and Inputs

The functions of the scope controls and inputs are summarized below. Numbers and letters refer to Fig. 5-1 for Tektronix T922 Oscilloscope or Fig. 5-2 for the Tektronix 2213 Oscilloscope.

1. INTENSITY controls brightness of the trace. Excessive intensity can burn the screen, especially if the spot is stationary. To avoid possible damage to the screen, turn down the intensity before turning the scope on.

2. FOCUS controls sharpness of the trace. Adjust for a fine, sharp trace.

3. BEAM FINDER aids in locating the trace when it is deflected off the screen. Pressing the Beam Finder button compresses the trace so that it can easily be centered using the position controls.

4. PROBE ADJUST provides a 1000 Hz square wave output of approximately 500 mV peak-to-peak, which is used when adjusting the probe compensation.

5A. CH1 VERTICAL POSITION moves the Channel 1 trace up and down.

5B. Same as 5A for Channel 2.

6. HORIZONTAL POSITION moves the trace sideways.

7A. CH1 VERTICAL SENSITIVITY (VOLTS/DIV) determines the CH1 vertical input voltage required to deflect the spot one division.

8A. CH1 VARIABLE VOLTS/DIV allows a smooth (rather than step-wise) variation of the CH1 vertical sensitivity. The marked values of VOLTS/DIV apply only when the red VARiable knob is in the calibrated position (rotated fully clockwise past the click stop).

8B. Same as 8A for CH2.

9A. CH1 AC-DC-GND input switch selects the coupling between the CH1 input and the vertical preamplifier. The AC position inserts a blocking capacitor so that only the AC component of the input is displayed. The DC position is direct coupled so that the entire input is displayed. DC should be used for low frequencies. The GND position grounds the amplifier input so the deflection is zero.

9B. Same as 9A except CH2 input is affected.

10A. CH1 VERTICAL INPUT permits connection of the circuit under test to the CH1 preamplifier. This connector is also the +INPUT for differential mode operation.

10B. CH2 VERTICAL INPUT permits connection of the circuit under test to the CH2 preamplifier. This connector is also the -INPUT for differential mode operation.

11. GROUND CONNECTOR is connected to the scope chassis (and building ground through the 3-wire line cord). The outer conductors on 10A, 10B and 19 are also grounds.

922 ONLY:

12. VERTICAL MODE buttons

CH1: Displays only the signal applied to the CH1 input.

CH2: Displays only the signal applied to the CH2 input.

DUAL TRACE: Displays both CH1 and CH2 input signals. Choice of ALTERNATE or CHOP dual-trace mode is automatically selected by the SEC/DIV setting. The CH1 input is used as the internal trigger source for both traces. The CH2 input may be selected as the trigger source by simultaneously depressing DUAL TRACE and CH2.

DIFFERENTIAL*: Selects the differential amplifier mode of operation with CH1 as the +INPUT and CH2 as the -INPUT. Internal trigger source is from CH1.

2213 ONLY:

12A. VERTICAL MODE switches

CH1: Displays only the signal applied to the CH1 input.

CH2: Displays only the signal applied to the CH2 input.

BOTH: Displays both the CH1 and CH2 inputs either separately or combined as determined by the settings of 12B and 12C.

12B. When 12A is set to BOTH:

ADD: Displays the sum of the CH1 and CH2 inputs.

ALTernate: Displays both the CH1 and CH2 inputs in the alternate dual-trace mode.

CHOP: Display both the CH1 and CH2 inputs in the chop dual-trace mode.

12C. INVERT inverts the CH2 input. When BOTH, ADD and INVERT are selected, the scope operates in the differential amplifier mode.

BOTH:

13. SEC/DIV selects the horizontal sweep rate, that is the time required for the spot to travel one division across the screen.

2213 ONLY: In the X-Y position, the sweep is disabled and the scope operates in the X-Y mode with CH1 as the X input and CH2 as the Y input.

---

*Some 922 scopes do not have this option.

922 ONLY:

14. X1-X10 provides calibrated sweep rates when in the X1 (fully counter-clockwise) click-stop position. Provides up to 10 times horizontal magnification when rotated clockwise.

2213 ONLY:

14. VARiable provides calibrated sweep rates when in the fully clockwise click-stop position. Increases the SEC/DIV up to 2.5 times when rotated counter-clockwise. Provides horizontal magnification by a factor of 10 when pulled out.

BOTH:

15. TRIGGERING SLOPE causes triggering to occur either on the positive (+) slope of triggering signal or on the negative (−) slope. (The SLOPE cannot be set to trigger on a zero slope.)

16. TRIGGERING LEVEL controls the voltage level of the triggering signal at which the sweep triggers.

17. TRIGGERING MODE switch −
    AUTO: Will provide automatic periodic triggering of the sweep if no triggering signal is present or if the maximum triggering signal amplitude is below that selected by the LEVEL control.

    NORM: Will trigger a sweep only if a trigger signal is present with the proper SLOPE and LEVEL.

    TV: Will provide proper synchronization with TV signals.

922 ONLY:

18. TRIGGERING SOURCE selects the signal used to trigger the sweep as follows:

    INT: internal trigger from the CH1 or CH2 signal as selected by the vertical mode

    LINE: trigger from the 60 Hz AC line

    EXT: trigger from the signal applied to the external (X) trigger input

    EXT/10: same as EXT but the external trigger signal is reduced by a factor of 10

    X−Y: disconnects the sweep signal from the horizontal amplifier and connects the external (X) horizontal input to provide horizontal deflection. The vertical (Y) signal is selected by the vertical mode button.

2213 ONLY:

18A. TRIGGERING SOURCE selects the signal used to trigger the sweep as follows:

    INT: internal trigger as selected by 18B

    LINE: trigger from the 60 Hz AC line

    EXT: trigger from the signal applied to the EXTernal INPUT

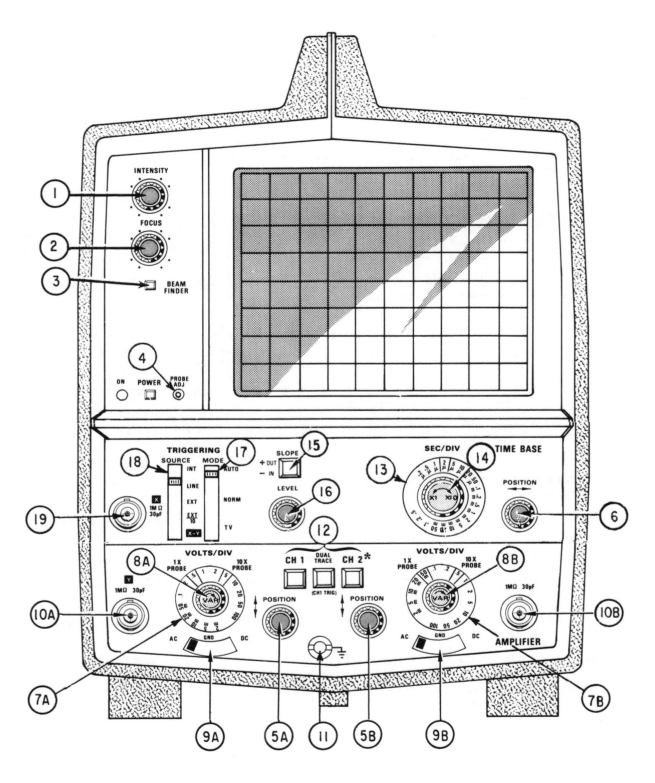

FIG. 5-1.   TEKTRONIX T922 OSCILLOSCOPE

Used by permission of Tektronix, Inc.

*Some T922 scopes have an optional 4th vertical mode button marked DIFF.

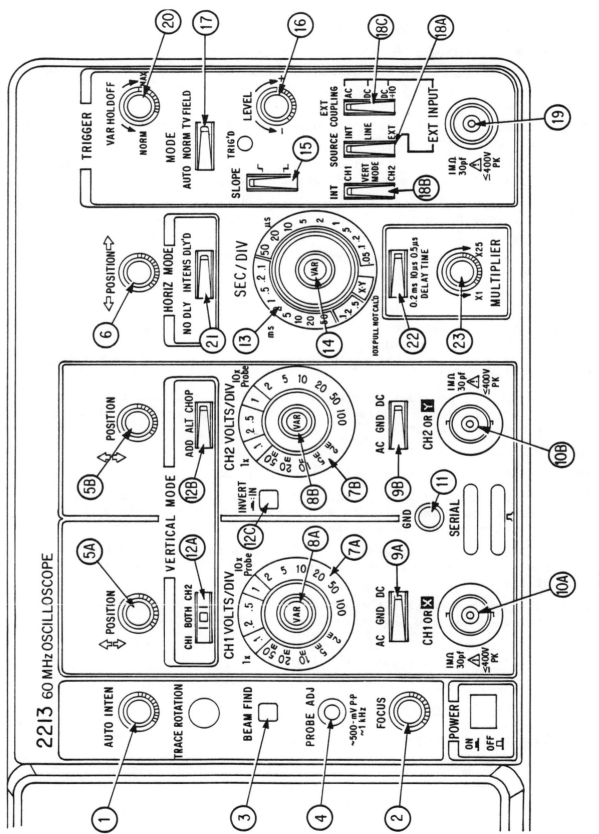

FIG. 5-2. TEKTRONIX 2213 OSCILLOSCOPE

Used by permission of Tektronix, Inc.

18B. INT selects the internal trigger source as follows:

CH1: trigger from the CH1 input signal

CH2: trigger from the CH2 input signal

VERT MODE: trigger from the displayed vertical signal except in CHOP and ADD; then the trigger is from the sum of the CH1 and CH2 signals

18C. EXT COUPLING selects the trigger coupling between an external triggering signal and the sweep circuits.

AC: a blocking capacitor is inserted so that only the AC component of the triggering signal affects the sweep

DC: the triggering signal is direct coupled so the entire signal (including DC component) controls the level at which the sweep triggers. DC coupling should be used for low frequencies.

DC÷10: same as DC except the external triggering signal is reduced by a factor of 10.

BOTH:

19. EXTernal INPUT is used for the external triggering signal when EXTernal trigger source is selected.

2213 ONLY:

20. VARiable HOLDOFF provides control of the holdoff time between sweeps. After a sweep has been completed, it cannot be retriggered until the holdoff time has elapsed.

21. HORIZ MODE switch –

NO DLY (delay): selects normal sweep operation

INTENSified: same as NO DLY except that a portion of the trace is intensified. The intensified zone starts after a delay, which is determined by the setting of 22 and 23.

DLY'D (delayed): the triggering of the sweep is delayed so that the sweep starts at the beginning of the INTENSified part of the trace.

22. DELAY TIME switch selects the amount of delay time between the start of the sweep and the beginning of the INTENsified zone.

23. MULTIPLIER increases the delay time selected by the DELAY TIME switch by a factor of approximately 1 to 25.

5.2 X-Y Operation of the Scope

The oscilloscope can be operated in two basic modes--the X-Y mode in which the input signals to both the horizontal and vertical amplifiers come from external sources, and the sweep mode in which the horizontal amplifier input is a sweep waveform generated within the scope. The X-Y mode is used to display one voltage against another, while the sweep mode is used to display a voltage as a function of time. In the X-Y mode, a positive voltage applied between the horizontal (X)

232

scope input and ground causes the spot on the screen to deflect to the right and a negative voltage to the left. The horizontal deflection in divisions is $x = v_h/S_h$ where v_h is the horizontal input voltage and S_h is the horizontal sensitivity in volts/div. A positive voltage applied between the vertical (Y) input and ground causes upward deflection of the spot and a negative voltage downward deflection. The vertical deflection in divisions is $y = v_v/S_v$ where v_v is the vertical input voltage and S_v is the vertical sensitivity.

If the signals applied to the vertical and horizontal inputs change slowly, the spot will move slowly on the screen. If the signals are periodic and the frequency is high enough, the spot will retrace the same path fast enough that a solid trace will appear on the screen, and no flicker will be observed because of persistence of the screen. If v_v and v_h are related by a constant, i.e., $v_v(t) = K\,v_h(t)$, the trace will be a diagonal line; if v_v and v_h are sine waves which are not in phase, the trace will be circular or elliptical.

5.3 Displaying Waveforms as a Function of Time

In order to dispaly a waveform as a function of time, the horizontal deflection must be proportional to time ($x = Kt$). The necessary horizontal signal is provided by the TIME BASE (see Fig. 5-3). Each time the TIME BASE is triggered, it produces a sweep waveform of the following form:

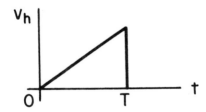

This waveform causes the spot to sweep across the screen once in T seconds and then return to the left edge of the screen. The rate at which the spot sweeps across the screen is determined by the TIME/DIV switch. If the sweep rate is R sec/div, the time required for the spot to travel X div is $t = R\,X$.

In order to display a steady picture on the screen, the sweep waveform must be repeated periodically, and it must be synchronized with the waveform being observed. This synchronization is accomplished by means of a triggering signal. This triggering signal may come from either vertical amplifier output (INTernal trigger), from the 60 Hz AC line (LINE trigger), or from the external trigger input (EXTernal trigger). Internal trigger is used when we want the triggering signal to be the same as the CH1 or CH2 vertical input, and external trigger is used when we want to observe waveforms with a time reference different from the CH1 and CH2 inputs. On many scopes, the triggering signal may be coupled to the time base

233

directly (DC coupling) or through a blocking capacitor (AC coupling). DC coupling is used when we want the entire signal (both AC and DC components) to affect the triggering circuit, while AC coupling is used when we want to control triggering with the AC component of the triggering signal. The Tektronix T922 scope is internally wired for AC trigger coupling in the external trigger mode and DC coupling in the internal trigger mode. The 2213 permits selection of AC or DC coupling in the external trigger mode.

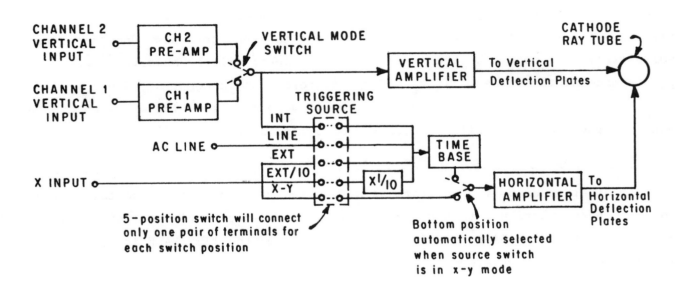

FIG. 5-3(A). SIMPLIFIED BLOCK DIAGRAM OF 922 SCOPE

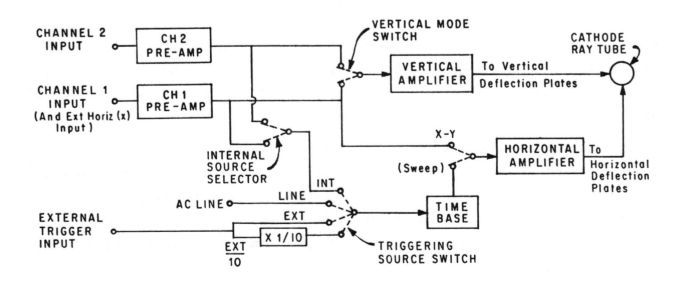

FIG. 5-3(B). SIMPLIFIED BLOCK DIAGRAM OF TYPICAL CALIBRATED X-Y SCOPE.

The TIME BASE is composed of a trigger pulse generator and a sweep generator (Fig. 5-4). A trigger pulse is produced when the slope and level of the triggering signal match the settings of the trigger SLOPE and LEVEL controls. Each trigger pulse will trigger the sweep generator and cause the beam to sweep across the screen once, provided that the previous sweep has gone to completion. Fig. 5-5 shows an example of trigger circuit operation.

In the AUTO trigger mode when no internal or external trigger signal is present, the sweep is automatically triggered by a signal from within the scope so that a base line will appear on the screen. The NORMal triggering mode requires an internal or external triggering signal of the appropriate LEVEL and SLOPE to trigger a sweep.

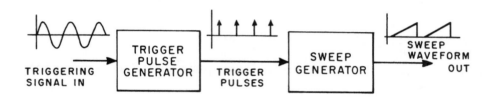

FIG. 5-4. OPERATION OF THE TIME BASE

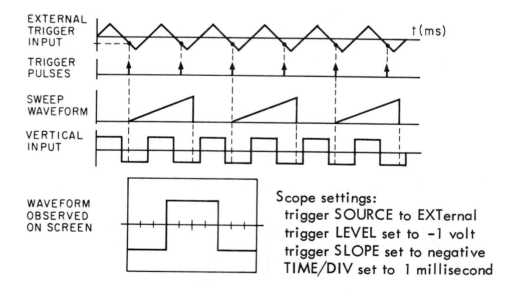

FIG. 5-5. EXAMPLE OF TRIGGER CIRCUIT OPERATION

Dual-trace operation allows two waveforms to be displayed simultaneously on the scope screen so that the relative timing of the waveforms may be observed. If CH1 is used as the trigger source, the CH1 display serves as the time reference, and the CH2 waveform is displayed with this same time reference. As shown in frame 2.53, an electronic switch alternately selects the CH1 and CH2 inputs for display. In the alternate sweep mode, during one sweep CH1 is displayed, during the next sweep CH2 is displayed, and so forth. For slow sweep rates, the alternate sweep mode causes excessive flicker, so the chopped mode is used. In the chopped mode, the electronic switch rapidly switches between the CH1 and CH2 inputs during each sweep. Since the electronic switch is not synchronized with the sweep and the switching frequency is high compared with the CH1 and CH2 input frequencies, the waveforms displayed on the screen appear to be continuous. For the T922 scope, alternate sweep is automatically selected for sweep rates faster than 1 ms/div and chopped sweep for slower sweep rates. In the dual-trace mode, the sweep may be triggered by either the CH1 or CH2 input.

5.4 Observing Signals with AC and DC Components

A periodic signal can have both an AC component and a DC component. Given a periodic waveform, $v(t)$, the average value is referred to as the DC component, V_{DC}. If we subtract out the DC component, the remainder is referred to as the time-varying or AC component, $v_{AC}(t)$; i.e.,

$$v_{AC}(t) = v(t) - V_{DC}$$

For the following waveform, the dashed line indicates the average or DC value and the AC component is plotted separately. Note that the average value of the AC component is zero with equal area above and below the line.

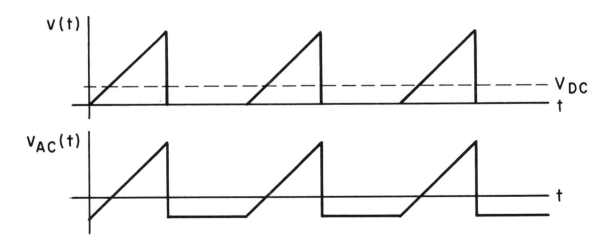

FIG. 5-6. WAVEFORM WITH AC AND DC COMPONENTS

When the input switch on the CH1 or CH2 vertical amplifier is set to AC, a blocking capacitor in series with the input removes the DC component so that only the AC component is observed. When the input switch is set to DC, the amplifier is direct coupled so that both the AC and DC components can be observed. The DC component of a waveform can be measured by switching the input from AC to DC and observing the vertical shift of the waveform. The DC input must be used for very low frequency signals since the impedance of the blocking capacitor $(1/\omega C)$ is not negligible at low frequencies.

5.5 The Differential Amplifier Mode

A differential amplifier amplifies the difference of the voltages applied to the +INPUT and -INPUT terminals. If v_+ is the voltage between the +INPUT and ground and v_- is the voltage between the -INPUT and ground, the difference component of the voltages is

$$\Delta v = v_+ - v_-$$

and the common-mode component is

$$v_c = \frac{1}{2} (v_+ + v_-)$$

An ideal differential amplifier amplifies the difference component and rejects the common-mode component, so the vertical deflection is

$$y = \frac{1}{S_v} (v_+ - v_-) = K_d \Delta v$$

where $K_d = 1/S_v$ is the difference-mode gain. If the amplifier is not perfect all of the common-mode component will not be rejected so

$$y = K_d \Delta v + K_c v_c = K_d (v_+ - v_-) + K_c \frac{v_+ + v_-}{2}$$

where K_c is the common-mode gain. The common-mode gain can be measured by applying a sine wave to both the +INPUT and -INPUT. With $v_+ = v_- = V_P \sin \omega t$ the observed deflection is

$$y = K_c V_P \sin \omega t = y_1 \sin \omega t$$

so $K_c = y_1/V_P$ div/volt. The common-mode rejection ratio is defined as

$$CMRR = K_d/K_c$$

An ideal differential amplifier has an infinite CMRR, and a good differential amplifier will have a CMRR greater than 100.

The error in the observed deflection may be significant if the common-mode signal is too large or if CMRR is too low. This error will be negligible if

$$|K_c v_c| \ll |K_d \Delta v| \qquad \text{or} \qquad \text{CMRR} = \frac{K_d}{K_c} \gg \left|\frac{v_c}{\Delta v}\right|$$

When the differential mode is used, both the CH1 and CH2 VOLTS/DIV switches must be set to the same value. The maximum input voltage which can be applied to either the CH1 or CH2 input in the differential mode is about 20 times the VOLTS/DIV switch setting. If this maximum is exceeded, the waveform may be distorted even though the difference of the v_+ and v_- inputs is small.

The differential mode is useful when neither terminal of the signal being observed can be grounded. It is also useful when measuring low-level signals in the presence of noise. For example, if

$$v_+ = v_{signal} + v_{noise} \qquad \text{and} \qquad v_- = v_{noise}$$

the vertical deflection will be

$$y = K_d(\Delta v + \frac{K_c}{K_d} v_c) = K_d(v_{signal} + \frac{v_{noise}}{\text{CMRR}})$$

so that most of the noise is rejected if CMRR is large.

5.6   Loading and Use of the Probe

The equivalent circuit of a typical scope input is a 1 megohm resistor in parallel with a 30 pf capacitor. When the scope input is connected to a high impedance circuit, the scope loads down the circuit so that the voltage being measured is reduced. The loading effect is worse at high frequencies because of the shunt capacitance in the scope input. The loading effect can be reduced by using a probe to increase the effective input impedance of the scope. A 10X attenuating probe (such as the Tektronix P6006 or P6108) attenuates (reduces) the input signal by a factor of 10, so when the probe is used the sensitivity switch setting must be multiplied by 10 to get the actual volts/div. The equivalent circuit for the probe and scope input is shown in Fig. 5-7.

238

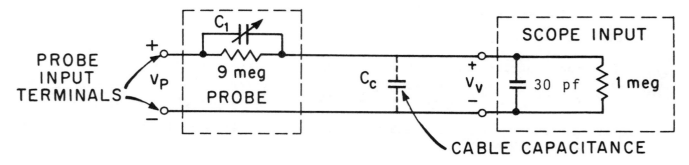

FIG. 5-7.   PROBE AND SCOPE INPUT

The capacitor in the probe must be properly adjusted so that the probe attenuation is independent of frequency. This can be accomplished by using the probe to observe a square wave, and adjusting the probe so that the corners of the square wave appear square with no overshoot or rounding (see Fig. 5-8).

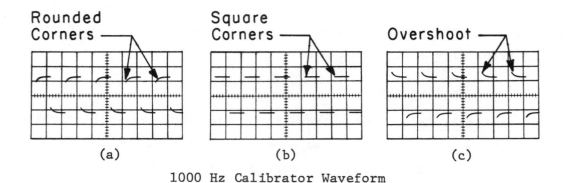

1000 Hz Calibrator Waveform

FIG. 5-8.   PROBE COMPENSATION

5.7 Making Accurate Measurements with the Scope

All of the "ground" terminals on the scope are connected together through the scope case and chassis, and they are connected to the building ground through the 3-wire line cord. Be careful not to short out part of your circuit by connecting two different points in the circuit to ground. If some point in the circuit is grounded, the circuit ground should be connected to the scope ground. (Do not rely on the line cord grounds for a connection).

When observing low amplitude signals, precautions are necessary to prevent excessive pickup of hum and noise. Keep your leads short and use a shielded cable or probe if necessary.

239

The following steps should be taken when voltage or time must be measured accurately with the scope:

(a) Use a fine, sharply focused trace. Avoid excessive intensity.

(b) Make sure the scope is properly calibrated.

(c) Expand the portion of the trace to be measured to fill as large a part of the screen as possible.

(d) Adjust the position controls to facilitate accurate reading.

(e) To avoid parallax error, view the screen from directly in front. (This is unnecessary if the scope has an internal graticule.)

(f) Use the probe if the circuit being measured has a high impedance or if the signals are high frequency.

(g) Make sure the frequency of the signals being observed is within the bandwidth of the scope (15 MHz for the 922 and 60 MHz for the 2213).

Even if all of the above precautions are taken in making measurements with the scope, the measured values may still have a calibration error as large as $\pm$ 3% and a reading error of $\pm$ 1/20 div or more.

5.8 Checking Scope Calibration

Vertical Amplifier Gain. The calibration accuracy of either vertical channel may be checked by measuring a DC voltage of known magnitude. Set the scope controls and adjust the DC input voltage to obtain a deflection of exactly 6 or 8 divisions. The VAR control must be in the calibrated position. Measure the DC voltage with a digital voltmeter. Calculate the expected deflection using this voltage and the VOLTS/DIV setting of the scope. Compare this expected deflection with the actual scope deflection to find the percent error of the scope:

$$\text{percent error} = \frac{\text{Actual scope deflection} - \text{expected deflection}}{\text{Expected deflection}} \times 100\,\%$$

If the percent error is larger than the manufacturers specifications for your scope (3% for the T922) then recalibration by a technician may be necessary.

CAL
X-Y
ONLY
Horizontal Amplifier Gain. The calibration of the horizontal amplifier may be checked using the same method as for the vertical amplifier. The scope must be placed in the X-Y MODE and the DC voltage applied to the external horizontal (X) input. The procedure is the same as above.

Time Base Accuracy. The calibration accuracy of the TIME BASE may be checked by using the scope to measure the period of a waveform of known frequency. A good choice for this waveform is the 60 Hz AC line stepped down through a transformer.

With the SEC/DIV set to 5 ms/div (calibrated), 3 cycles of the 60 Hz sinewave should produce exactly 10 divisions deflection. The percent error may be calculated by comparing the actual deflection to the expected 10 divisions using the formula above. If the percent error exceeds specifications (3% for the 922) then recalibration may be necessary.

5.9 Measurement of Phase Angle

Given two sinusoidal voltages $v_1(t) = A \sin \omega t$ and $v_2(t) = B \sin(\omega t + \theta)$, the phase angle θ can be determined by the dual-trace method, the triggered sweep method, or the ellipse method. The first two methods determine both sign and magnitude of θ, and the latter determines only the magnitude.

Dual Trace Method. First connect v_1 to the CH1 vertical input. Using the CH1 internal triggering mode, display $v_1(t)$ and calibrate the time axis to 20°/div (or some other convenient value):

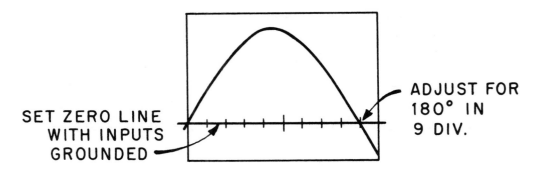

This establishes a t = 0 reference point. Connect v_2 to the CH2 input, and select the dual trace mode. Still using CH1 as the trigger source, and without changing any of the triggering controls, observe the phase of v_2 relative to v_1. If $v_2(0)$ is positive, $\theta = 180° - \theta_2$ where θ_2 is the distance between the origin and the "180° point" on the sine wave. If $v_2(0)$ is negative, then $\theta = -\theta_1$.

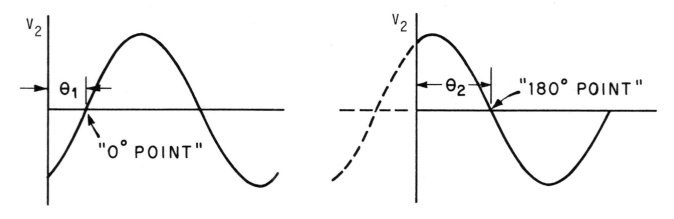

FIG. 5-9.   PHASE ANGLE MEASUREMENT

241

<u>Triggered Sweep Method</u>. If your scope does not have dual trace operation, the triggered sweep method can be used to determine the magnitude and sign of the phase shift. This method is also useful when v_1 and v_2 do not have a common ground so that differential mode operation must be used. The triggered sweep method is similar to the dual-trace method, except the external trigger must be used and only one waveform is observed at a time. To calibrate the time axis, connect v_1 to the CH1 vertical input and to the EXTernal trigger input, and use EXTernal trigger mode. Once the time axis is calibrated, do not disturb the trigger circuit (leave v_1 connected to EXTernal trigger). Connect v_2 to the CH1 input (or to CH1 and CH2 if differential mode operation is required) and observe v_2. Read the phase angle as in the dual-trace method.

<u>Ellipse Method</u>. Connect v_1 to the horizontal (X) input, v_2 to the vertical (Y) input, and display v_2 vs v_1 on the screen. The magnitude of θ can be determined from Table 5-1, or it can be read directly off the trace by using a Webb mask (Fig. 4-49). The ellipse method does not give the sign of θ; however, if the sign is known at one frequency and the ellipse is observed as the frequency is varied continuously, a change of sign can be detected when the ellipse closes to a straight line and opens out again. For accurate results, the ellipse method requires that the phase shift introduced by the vertical and horizontal scope amplifiers be the same; otherwise, phase shift compensation is necessary.

Table 5-2. Determination of Phase Angle by the Ellipse Method

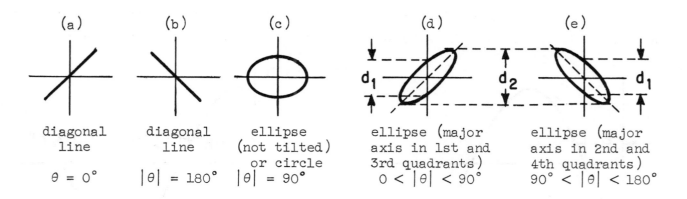

For (d) and (e), $|\theta| = \sin^{-1}(d_1/d_2)$ where d_1 is the distance between the y-axis intercepts and d_2 is the peak-to-peak vertical deflection. The ellipse must be centered horizontally.

<u>Lissajous Figures.</u> A Lissajous figure can be displayed on the scope by applying sine waves of different frequencies to the X and Y inputs. If the ratio of the frequencies of the two sine waves reduces to a ratio of small integers, a stable Lissajous figure may be observed. If the two sine waves are not precisely synchronized, the figure will slowly drift from fully open to fully closed again. The fully open figure is like a twisted loop with no end points, while the fully closed figure is like a twisted line with definite endpoints (see Fig. 5-10). The ratio of the vertical input frequency (f_v) to the horizontal input frequency (f_h) can easily be determined from a fully open Lissajous figure as

$$\frac{f_v}{f_h} = \frac{\text{number of vertical maximums}}{\text{number of horizontal maximums}}$$

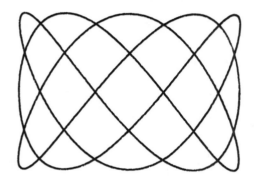

    (a) Fully open   (b) Fully closed

FIG. 5-10. LISSAJOUS FIGURES WITH $f_v/f_h = 5/3$

The following equipment is required for the laboratory parts of the text.

     triggered sweep oscilloscope (see Appendix B)

     dual low-voltage DC power supply (or single supplies) capable of supplying 0 to $\pm$ 20 volts at $\geq$ 500 mA

     Digital voltmeter, DC and AC (RMS for sine input)

     function generator for sine, square and triangular waves with a frequency range of at least 0.1 Hz to 50 KHz and amplitude $\geq$ 20 volts peak to peak

     sine wave oscillator (or a second function generator) with a frequency range of at least 20Hz to 40KHz

     three 3-foot coaxial cable leads, each with a BNC connector to fit scope inputs on one end and red and black clips on the other end

     additional leads as necessary to connect to above equipment

Lab Part I - additional equipment required:

     R-C network

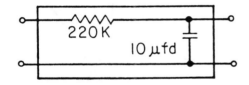

     continuity tester

Lab Part II - additional equipment required

     2200 ohm resistor

     0.1 microfarad capacitor

Lab Part III - additional equipment required

     10X attenuating probe

     2 500 kilohm 1% precision resistors

     1K, 2.7K, 5.6K, 10K, 100K and 1 megohm 10% resistors

     0.1 microfarad capacitor

     adaptor to convert 3-prong grounding plug to 2-prong plug

Lab Part IV - additional equipment required

     Active phase shift network (see notes below)

     Webb phase shift mask (optional)

Fig A-1 shows the circuit diagram for the active phase shift network. Construct the network using the components listed below.

Parts List

R_1, R_2	5.1 kilohm, matched within 1%
R_3, R_4	5.1 kilohm, matched within 1%
R_5	1.8 kilohm $\pm$ 5%
R_6	500 ohm 10 turn potentiometer (Trimpot 3006F-1501)
R_7	1 kilohm $\pm$ 5%
R_8	2 kilohm $\pm$ 5%
C_1, C_2	0.05 µfd $\pm$ 5%
C_3, C_4	0.1 µfd

R_6 should be adjusted for 180° phase shift (v_{34} relative to v_{12}) at 1590 Hz.

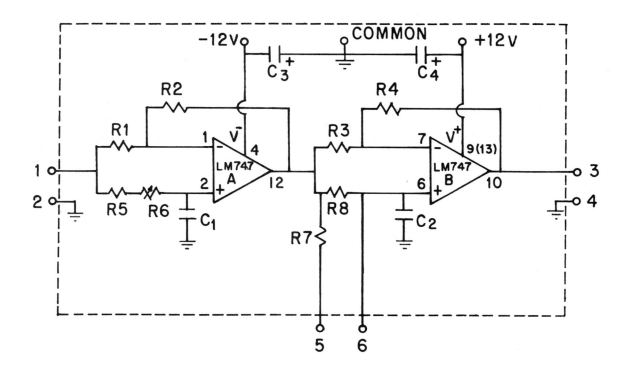

FIG. A-1. CIRCUIT DIAGRAM FOR ACTIVE PHASE SHIFT NETWORK

APPENDIX B - SUITABLE OSCILLOSCOPES

   Table B-1 gives a list of oscilloscopes which are suitable for use with the text. This list is not intended to be exclusive, since the lab exercises in this text can be adapted for use with almost any triggered-sweep oscilloscope. Your lab instructor should provide a list of any changes to the text which are necessary for the particular type of scope you are using.

   This text is designed to work with two different categories of oscilloscopes. The primary difference between these two categories is in the X-Y mode of operation. One type of oscilloscope has <u>calibrated</u> VOLTS/DIV switches for both the X and Y channels. Instructions in the text which apply only to this type of scope are labeled "CAL X-Y ONLY". The other type of oscilloscope has a calibrated VOLTS/DIV switch only on the Y channel. Instructions in the text which apply only to this type of scope are labeled "922 ONLY", since the Tektronix 922 is a typical scope in this category. Table B-1 indicates whether you should follow the "CAL X-Y ONLY" or "922 ONLY" instructions for your oscilloscope.

Table B-1. Suitable Oscilloscopes

model	use instructions for	remarks
Tektronix 921 (15MHz)	922	omit parts on dual-trace and differential mode
922, 922R (15MHz)	922	
932A, 935A (35MHz)	CAL X-Y	has additional controls not discussed in text
2213, 2215 (60MHz)	CAL X-Y	
B&K Dynascan 1477 (15MHz), 1479 (30MHz)	CAL X-Y	use CHA-CHB instead of CH1-CH2; CHB provides horizontal deflection in X-Y mode
Hewlett-Packard 1220A (15MHz)	922	omit parts on dual-trace and differential mode
1222A (15MHz)	CAL X-Y	use CHA-CHB instead of CH1-CH2
Hitachi V-151 (15MHz), V-301 (30MHz)	922	omit parts on dual-trace and differential mode
V-152 (15MHz), V-302 (30MHz)	CAL X-Y	
Gould OS255 (15MHz)	CAL X-Y	CH2 provides horizontal deflection in X-Y mode
Heathkit IO-4510 (15MHz)	CAL X-Y	
IO-4535 (35MHz)	CAL X-Y	has additional controls not discussed in text
Kikusui 5520 (20MHz)	CAL X-Y	has additional controls not discussed in text
5530 (35MHz)	CAL X-Y	